T0304554

Electronic Interfaces for Differential Capacitive Sensors

RIVER PUBLISHERS SERIES IN ELECTRONIC MATERIALS AND DEVICES

Series Editors

EDOARDO CHARBON
EPFL
Switzerland

MIKAEL ÖSTLING
KTH Stockholm
Sweden

ALBERT WANG
University of California
Riverside, USA

Indexing: All books published in this series are submitted to the Web of Science Book Citation Index (BkCI), to SCOPUS, to CrossRef and to Google Scholar for evaluation and indexing.

The "River Publishers Series in Electronic Materials and Devices" is a series of comprehensive academic and professional books which focus on the theory and applications of advanced electronic materials and devices. The series focuses on topics ranging from the theory, modeling, devices, performance and reliability of electron and ion integrated circuit devices and interconnects, insulators, metals, organic materials, micro-plasmas, semiconductors, quantum-effect structures, vacuum devices, and emerging materials. Applications of devices in biomedical electronics, computation, communications, displays, MEMS, imaging, micro-actuators, nanoelectronics, optoelectronics, photovoltaics, power ICs and micro-sensors are also covered.

Books published in the series include research monographs, edited volumes, handbooks and textbooks. The books provide professionals, researchers, educators, and advanced students in the field with an invaluable insight into the latest research and developments.

Topics covered in the series include, but are by no means restricted to the following:

- Integrated circuit devices
- Interconnects
- Insulators
- Organic materials
- Semiconductors
- Qantum-effect structures
- Vacuum devices
- Biomedical electronics
- Displays and imaging
- MEMS
- Sensors and actuators
- Nanoelectronics
- Optoelectronics
- Photovoltaics
- Power ICs

For a list of other books in this series, visit www.riverpublishers.com

Electronic Interfaces for Differential Capacitive Sensors

Gianluca Barile

Giuseppe Ferri

Vincenzo Stornelli

Università degli Studi dell'Aquila, Italy

LONDON AND NEW YORK

Published 2020 by River Publishers
River Publishers
Alsbjergvej 10, 9260 Gistrup, Denmark
www.riverpublishers.com

Distributed exclusively by Routledge
4 Park Square, Milton Park, Abingdon, Oxon OX14 4RN
605 Third Avenue, New York, NY 10017, USA

Electronic Interfaces for Differential Capacitive Sensors / by Gianluca Barile, Giuseppe Ferri, Vincenzo Stornelli.

Routledge is an imprint of the Taylor & Francis Group, an informa business

ISBN 978-87-7022-150-4 (print)

While every effort is made to provide dependable information, the publisher, authors, and editors cannot be held responsible for any errors or omissions.

Contents

Preface

In a world where integrated and highly autonomous systems are adopted in many areas of everyday life, great efforts are spent in the research and industrial environments for the design of electronics systems able to harvest energy from the environment around them and, in general, to maintain a very high efficiency level across all their possible working conditions. As a matter of fact, measurement subsystems (which are ubiquitous in any application field) play a significant role in this regard since, on the one hand, the transducer almost inevitably needs a power source to convert the measurand, and, on the other hand, being the first electronic interface necessarily analog, it does not gain benefits, in terms of power consumption, from the simple technological scaling as, on the contrary, does the processing part of the complex system, typically digital. That said, on the one hand, it is necessary to use transducers which, by their nature, guarantee a very low power consumption, and, on the other, it is worth carrying out research aimed at the design of interface circuits that maximize readout performances optimizing the aforementioned constraints. In light of this, capacitive sensors have established in many applications, replacing the piezoresistive counterparts, since they consume virtually zero in terms of dissipated power, while not giving up in terms of robustness to temperature variations and sensitivity.

This book focuses on showing to the reader the state of the art and the scientific developments of new circuits and systems employed in the readout of differential capacitive sensors. Being a subset of the capacitive ones, this type of sensors has their same advantages while being inherently excellent at rejecting common mode disturbances. Combining the aforementioned features, together with the possibility to micromachine them directly onto a silicon substrate (hence together with the electronic readout circuit), makes them very appealing in many applications where dimensional scaling is a real necessity.

The book is divided into chapters as follows. The first chapter will be focused on a general introduction to the sensor world, both from the transducer and from the electronic interface points of view. After a brief review of sensors, in the first part the main parameters typically used to evaluate static and dynamic performances of these devices will be presented to the reader. In the second part, we will introduce the most common electronic circuits adopted to suitably evaluate the possible variations of a transducer. To better highlight the features of each configuration, we will offer some examples that make use of resistive sensors as generic test benches. Lastly, we will cite a number of techniques to increase the accuracy of the interface during the measurement.

The second chapter will introduce capacitive transducers and how to perform their readout through suitable electronic circuits. Standard and differential capacitive sensors will be analyzed, characterizing them from a physical and circuital point of view. The corresponding analytical parameterization of this type of sensors is also treated, based on its physical nature; insights on the advantages and disadvantages of various alternatives will be presented. Along the text, electronic circuits for simple and differential capacitive sensors will be shown. In the final section of the chapter, an analysis on the effects of parasitic components of a differential capacitive sensor will also be performed.

In the third chapter, we will analyze the state of the art and recent advances on interfaces and active basic blocks for differential capacitive sensors. In particular, this chapter is focused on the voltage mode approach. Indeed, there are several alternatives that fall into this category. The most common is represented by switched capacitor architectures whose operations depend on the status of a number of internal switches. These interfaces, although effective, suffer from various relevant problems related to the nonideality of the switches (clock feedthrough and charge injection). In this sense, we will also show "switchless" interfaces, typically based on oscillators, capable of adjusting the frequency (or duty cycle) of the output signal according to the value of the transducer. The limit of these proposals is linked, in particular, to the stability of the oscillator itself and to the almost inevitable mismatch between the passive components that constitute the circuit. Other alternatives are also analyzed, in particular, the differential capacitance to voltage conversion, and also alternatives that showcase a direct digital conversion.

The fourth chapter follows the same pattern as the previous one, but focusing on all these interfaces that use a current mode approach. Unlike the voltage mode, the literature, in this topic, is rather standardized; in other words, all proposals are based on a well-defined theory. Although, indeed, encapsulating information in a current has numerous advantages over a purely voltage mode, making it possible to obtain extremely simple architectures, allowing high-speed readings and high-sensitivity values, at a very low power consumption, has also crucial drawbacks. In fact, a current mode interface results extremely sensitive to the nonidealities of the sensor and of the electronics itself (parasitic capacitances and offset of the amplifiers). Moreover, the voltage across a capacitor, under the action of a constant reference current, tends to linearly increase, making it necessary for the designer to find a way to periodically discharge it. The chapter will be divided into a first section where all the interfaces designed with "standard active devices" such as Op-Amps and OTAs will be discussed. In the second part, the focus will be directed to the interfaces implemented by means of "nonstandard devices," namely second-generation current and voltage conveyors (CCIIs and VCIIs).

In the fifth chapter, we will show how the concept of autobalanced bridge has been utilized to interface differential capacitive sensors. The first part of the chapter will introduce the main aspects of this technique and how it is possible to use a synchronous demodulation-based feedback to increase the readout accuracy. Both linear and nonlinear interfaces based on this approach will be presented, together with related circuits and systems employed to mitigate the effects of parasitic capacitances in this scenario.

In order to conclude the book, and to allow the reader to have a better understanding of the working principle of some of the previously proposed circuits, we offer an appendix where nonstandard active blocks that can be profitably employed as building elements for interface circuits and systems are recalled. In particular, in the first half of the appendix, we will introduce the second-generation current conveyor, CCII, its ideal relationship, the nonidealities and some transistor-level topologies based on the differential pair architecture. Similarly, in the second part, we will repeat the same analysis for the second-generation voltage conveyor, VCII. External nodes relationships will be given, and topologies to synthesize the aforementioned VCII will be finally analyzed.

List of Figures

List of Abbreviations

AC   Alternate Current
ADC   Analog to Digital Converter
CCI   First Generation Current Conveyor
CCII   Second Generation Current Conveyor
CDS   Correlated Double Sensing
CMOS   Complementary Metal Oxide Semiconductor
CSA   Charge Sensitive Amplifier
DC   Direct Current
DR   Dynamic Range
INA   INstrumentation Amplifier
LSB   Least Significant Bit
MEMS   Microelectromechanical System
MOS   Metal Oxide Semiconductor
MOSFET   Metal Oxide Semiconductor Field Effect Transistor
NIC   Negative Impedance Converter
Op-Amp   Operational Amplifier
OTA   Operational Transconductance Amplifier
TIA   Transimpedance Amplifier
VCI   Voltage Controlled Impedance
VCII   Second Generation Voltage Conveyor
VCNIC   Voltage Controlled Negative Impedance Converter
VCR   Voltage Controlled Resistance

1

Introduction on Sensor Systems: Transducers, Sensors, and Electronic Interfaces

From the most common areas, such as monitoring of industrial process parameters, to the newest applications, such as internet of things (IoT) and self-driving vehicles; from the medical sector, with the need of monitoring vital parameters in the least invasive way possible, to the space applications where reliable, accurate, and efficient data collection is required, the correct functioning of the overall system depends on those electronic devices designed to perceive variations in the mechanical, chemical, hydraulic, and electromagnetic magnitudes under examination.

1.1 Sensors: A Brief Review

Generally speaking, the name given to devices able to detect real phenomena is "transducers." A transducer is a device capable of converting a magnitude, often called measurand, from an energy domain to another, which does not necessarily correspond to the electrical one. Signal processing, however, needs electrical inputs to be acquired; this means that the energy domain related to a "sensing" process is always the electrical one.

Devices or systems, which convert energy from a given domain to the electrical one, are called "sensors." A sensor either can correspond to a single transducer (see Figure 1.1(a)) or can be seen as a complex system made of one or more transducers, followed by an electronic circuit that carries out the final conversion (see Figure 1.1(b)). For instance, a piezoelectric device is both a transducer and a sensor, since it is capable of converting vibrations into an electrical signal. On the other hand, a hot wire anemometer, like the name suggests, evaluates the wind speed by measuring how much a heater gets cooled down from a starting temperature. Therefore, the wind speed

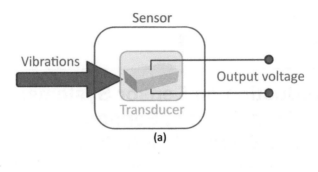

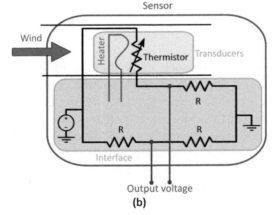

Figure 1.1   (a) A single transducer sensor and (b) a multiple transducers plus sensor interface.

is converted into heat, which in turn varies a thermistor resistance whose changes are sent to an electronic readout interface (a Wheatstone bridge in Figure 1.1(b)).

The information coming from a sensor (or a sensing system) can be of different nature. As reported in the previous figures, it can be fully analog, but also digital, ready to be processed by a microcontroller. In particular, in the latter case, analog-to-digital converters are required. Moreover, it is often necessary to equip the overall system with dual devices, that is, digital-to-analog converters that supply one or more actuators, which carry out an inverse transduction, from an electrical domain to a nonelectric one (the "actuation").

We can classify sensors in several ways. With respect to the power supply, they can be active or passive. Active sensors are all these devices that require an additional external signal for their use. Possible examples are resistive and capacitive sensors, where some kind of voltage has to be

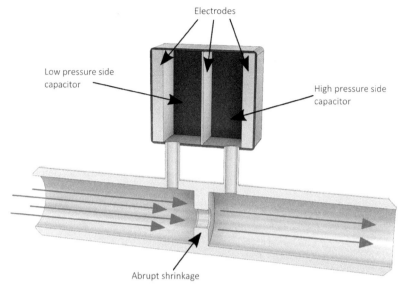

Figure 1.2   An example of differential capacitive flow sensor.

applied to investigate their variation under the action of a measurand. On the contrary, passive sensors do not require external signals since they are inherently capable of outputting an electrical signal. As an example, again, a piezoelectric sensor falls into this category. Another possible classification is based on the measurand type. In this sense, we can consider basically four categories: hydraulic sensors, mechanical sensors, environmental sensors, and electromagnetic sensors [1–3].

Typically, pressure and flow sensors are the most well-known types that fall into the hydraulic category. Figure 1.2 represents an example of a hydraulic capacitive sensor. Also there are several cutting-edge research fields where these sensors are used such as microfluidics and lab-on-chip devices where a very small amount of liquid needs to be properly driven. Mechanical sensors are widespread since they are sensitive to changes in mechanical properties. As an example, Figure 1.3(a) depicts a simple uniaxial resistive extensometer, while in Figure 1.3(b), a triaxial MEMS environmental accelerometer is shown. Environmental monitoring consists of evaluating a set of measurands which can assess the overall status of a given environment.

Some indicative application fields of these kind of sensors may be the tracking of metals into samples (nanoelectrode arrays), the identification of one or more substances into a gas (chemical sensor arrays), and the

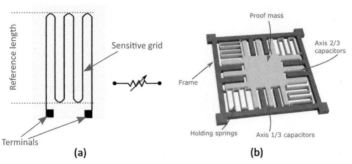

Figure 1.3   (a) A uniaxial resistive extensometer and (b) a triaxial capacitive accelerometer.

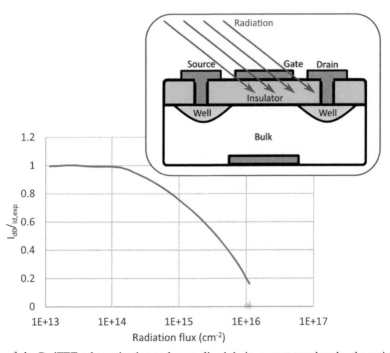

Figure 1.4   RadFET schematization and normalized drain current trend under the action of radiation flux.

evaluation of radiation flux levels (RadFETs, radiation field-effect transistors, revealing the radiation flux through the determination of a current, see Figure 1.4) [4–6].

Lastly, electromagnetic sensors are typically utilized when a wireless and contactless measurement has to be carried out. Magnetic sensing generally

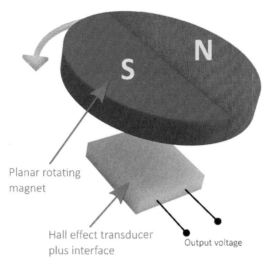

Planar rotating
magnet

Hall effect transducer
plus interface

Output voltage

Figure 1.5 A magnetic encoder: a Hall effect sensor evaluates the variation in the magnetic field induced by the rotation of a planar magnet.

reveals the magnetic field through suitable materials that generate a somehow variable field. Figure 1.5 depicts this framework showing a magnetic encoder which, thanks to a Hall effect sensor and a rotating magnet, allows to measure an angle or an angular velocity [7, 8].

1.2 Sensor Parameters Overview

Regardless of the complexity of the sensor (may it be a single transducer or include also a very complex interfacing circuitry), a set of parameters can be identified so to describe its behavior under different working conditions. Specifically, it is possible to define two categories of parameters: static and dynamic. The former describes the behavior of the sensor at the steady state. It indicates both when there is no measurand (baseline conditions of the sensor) to be detected and when there is a static magnitude (or slowly variable) to be evaluated. The latter category describes how fast and accurately the sensor transitions occur from a steady state to another.

1.2.1 Static Parameters

The theoretical relationship that links a measurand variation to the corresponding electrical output variation of the sensor is generally called *transfer function (or input–output relationship).* Each point of the input–output relationship describes the behavior of the sensor for a given measurand input at the steady state (static conditions). Very often, the designer aims to obtain a unidimensional and linear input–output relationship. This means that, ideally, on the one hand, the output of the sensor is insensitive to any magnitude different from the one that has to be evaluated and, on the other hand, that the slope of the tangent line at any given point is a constant value. This magnitude, in particular, is defined as *sensitivity* and represents the rate at which the output varies for a given input change. It can be expressed as:

$$S = \frac{dOut\,(x_0)}{dx} \tag{1.1}$$

where x is the measurand, x_0 is a generic value of the measurand where the derivative is calculated, and *Out* is the electric output magnitude.

There are cases when the transfer function is nonlinear as depicted in Figure 1.6(a). In this scenario, the shape of a typical transfer function is given by Equation (1.2):

$$Out = \frac{a + bx + c(1 + x)}{d(1 + x)} \tag{1.2}$$

In the following figures, the measurand x is evaluated by means of absolute variation (expressed as a percentage) with respect to the baseline value.

In these conditions, the sensitivity (Figure 1.6(b)) is a variable, which means that its value depends on the actual measurand magnitude (X_0). This is unideal because, as stressed by figures below, the sensor is very sensitive to small measurand values, but it becomes almost insensitive to higher ones. In this situation, the nonlinear curve is often approximated by line pieces for different measurand magnitude ranges (piecewise approximation).

A high sensitivity is important to achieve, because it directly influences the *resolution* (R) of the sensor. This parameter identifies the minimum variation of the measurand that produces a detectable variation of the sensor output. In other words, it determines the minimum variation that the sensor is able to perceive with no uncertainty. In case of a fully analog output, it is possible to define this magnitude as:

$$R = \frac{Noise}{S} \tag{1.3}$$

The unit of measurement of *R* is the same as the measurand one. The noise parameter indicates the *output noise*, which has to be considered according

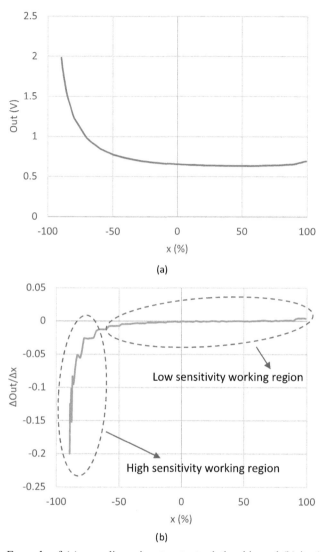

Figure 1.6 Example of (a) a nonlinear input–output relationship and (b) its (nonconstant) sensitivity using the following coefficients in Equation (1.2): a = 6.62, b = 4.62, c = 1.21, d = 12.1.

to the electrical output type. In the case of a digital output, the resolution is often given as:

$$R = \frac{Value}{LSB} \qquad (1.4)$$

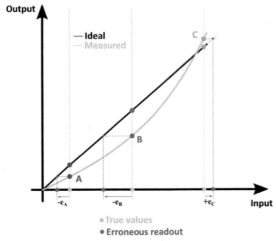

Figure 1.7   Accuracy evaluation for a real-world transfer function.

where *value* represents the minimum amount of measurand variation that a *least significant bit* (LSB) variation represents.

Considering the ideal versus real transfer function shown in Figure 1.7, it is possible to introduce two more parameters, *accuracy* and *precision*. The accuracy of a sensor is defined as the absolute maximum error in evaluating the measurand that can be experienced over the entire input–output relationship curve. Referring to Figure 1.7, errors at each point can be defined as the difference between the expected measurand value (which is obtained substituting the measured output magnitude into the ideal input–output relationship, red dots on the input axis) and the true value (green dots on the input axis). The maximum value among all these errors is indeed defined as *accuracy* ($-e_B$ in Figure 1.7). Since that, by repeating the measurement of the same input, the output readout will always be slightly different, the accuracy at any given input can be then defined as the mean value of the probability density function obtained by the measurements. Analogously, the standard deviation of the probability function is the *precision* of the interface at that specific input value. Figure 1.8 represents this statement. An accurate sensor grants that, by mediating a certain number of measurements, the measured value is close enough to the true one. An accurate and precise sensor grants that, by executing a single measurement, this value is very close to the true one.

It is important to underline that what it has been said stands for an uncalibrated sensor. This means that the real-world input–output relationship is unknown. If the accuracy does not satisfy the application requirements, it

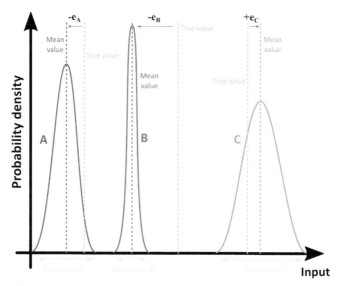

Figure 1.8 Accuracy and precision over three measurement sets.

is necessary to calibrate it, so to measure the actual input–output relationship of the given sensor and ideally to nullify its inaccuracy.

If the sensor has a linear ideal transfer function, it is possible to define the so-called linearity error as the absolute maximum distance between the real transfer function and the straight line that approximates it. It can be evaluated as a percentage value referred to the full-scale input or in terms of input magnitude. There are several techniques to trace the approximating line. The most often used methods are the *end-point method* and the *best-fit straight-line method*. The former is the simplest one because it consists of tracing the line that contains the two ending points of the input–output characteristic (Figure 1.9(a)). This technique gives the maximum possible linearity error. The latter method is a statistical one. It is based on the measurement of a sufficiently high number of points of the input–output characteristic over the full-scale input range. Applying the least square method on these points, it is possible to calculate the slope and the intercept of the best fitting line (Figure 1.9(b)).

The parameter called *hysteresis* indicates the difference between the readout results of a given measurand value if its value is determined from the lower side of the input–output characteristic rather than from the upper side. In other words, the hysteresis of a measured transfer function indicates the input-referred difference between the transfer function obtained sweeping

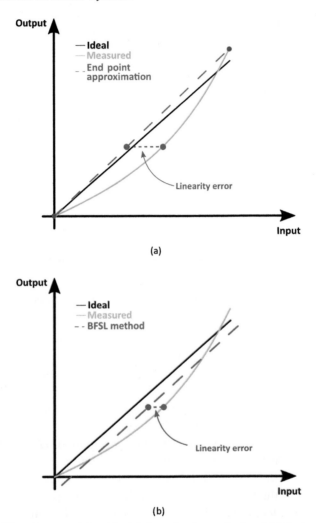

Figure 1.9   Linearity error defined by (a) the end-point method and (b) the best-fit straight-line method.

the measurand from the lower values to the higher values and the one that is obtained sweeping the measurand from the higher values to the lower values (see Figure 1.10).

The parameter that indicates the span of the measurand that the sensor is effectively capable of reading is called *dynamic input range*. It can be expressed according to the unit of measurement of the input magnitude. The

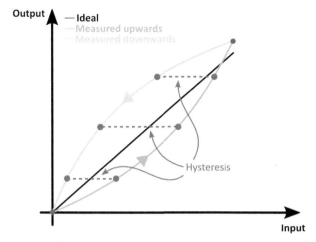

Figure 1.10   Hysteresis effects on an input–output relationship.

boundaries of the input range are chosen considering the maximum accuracy error allowable by the application.

The maximum input value for the measurand is called full-scale input, which at the output of the sensor generates the full-scale value, correspondingly.

1.2.2 Dynamic Parameters

As previously stated, the dynamic performances of a sensor are related to its transient behavior between a steady state and the other. Indeed, all sensors have a finite response time to an instantaneous change in physical signal. This means that, if the variations of the measurand follow each other too quickly and the sensor cannot keep up with them, even in case of an ideal input–output relationship, the readout process would end up with an error called *dynamic error* (Figure 1.11).

Based on the dynamic behavior, a sensor can be characterized by its order. The order of a sensor basically indicates the order of the linear differential equation that models the dynamic response of the system as a function of time. More in general, it is possible to mainly identify zero-order, first-order, and second-order sensors.

Higher-order differential equations are typically unused. From a practical point of view, this parameter indicates the "inertial" elements contained in the sensor. For example, it can be a proof mass, a spring, or a capacitor.

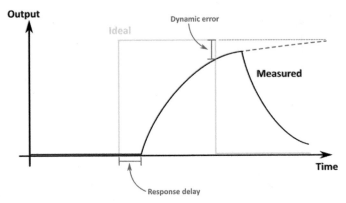

Figure 1.11   Readout error caused by the speed of the sensor.

A zero-order sensor does not need any specific dynamic analysis since it is a system that does not contain inertial elements. It has no dynamic error and infinite bandwidth, and it can be described by the following equation:

$$Out(t) = kIn(t) \tag{1.5}$$

Such a sensor is purely ideal and a system hardly behaves like this in real world. In fact, for instance, even resistive transducers tend to have some kind of parasitic capacitance that acts like an inertial element. In other words, assuming a sensor that behaves as a zero-order system, this is valid only in specific conditions and only for very small variations.

First-order sensors are characterized by the following equations, in time and Laplace domain, respectively:

$$In(t) = \frac{k_1 \, dOut(t)}{dt} + k_0 \, Out(t)$$

$$F(s) = \frac{Out(s)}{In(s)} = \frac{S}{\tau s + 1} \tag{1.6}$$

They can be thought as systems that have an element that stores energy and one that dissipates it. From the first-order up, a sensor can be characterized both in the frequency domain (Bode plots) and in the time domain (step response). Bode plots for a first-order sensor are shown in Figure 1.12 (upper plot: magnitude, lower plot: phase). In Equation (1.6), S represents the static sensitivity of the sensor. We can define the cutoff frequency f_c as the frequency where the gain drops by 3 dB and the phase shift reaches $-45°$.

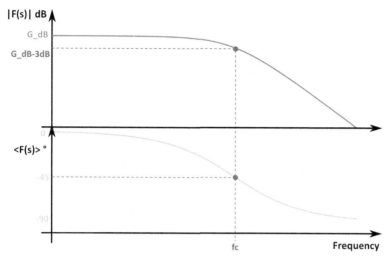

Figure 1.12   Bode plots for a first-order sensor.

It is possible to correlate the parameters in Equation (1.6) as follows:

$$S = \frac{1}{k_0}$$

$$f_c = \frac{\omega_c}{2\pi} = \frac{1}{2\pi\tau} = \frac{k_0}{2\pi k_1}$$

(1.7)

The cutoff frequency *fc* indicates the upper boundary for the input measurand frequency. Similarly, a lower cutoff frequency can exist. The parameter τ is often used for evaluating how fast the sensor reaches the steady state in terms of time domain and is called *time constant* (typically, for t > 3τ, the sensor output is considered at the steady state). Figure 1.13 shows the step response of a first-order sensor. It is possible to define *speed response* as the amount of time that the output takes to go from 10% to 90% of the final value, for a unitary measurand step variation.

Second-order sensors are characterized by the following equations:

$$In(t) = \frac{k_2 d^2 Out(t)}{dt^2} + \frac{k_1 d Out(t)}{dt} + k_0 Out(t)$$

$$F(s) = \frac{Out(s)}{In(s)} = \frac{S\omega_n^2}{s^2 + 2\zeta\omega_n s + \omega_n^2}$$

(1.8)

S is once more the static sensitivity of the sensor, while ω_n is called natural frequency and represents the value where, if undamped, the magnitude

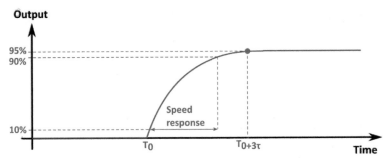

Figure 1.13   Time response of a first-order sensor when a step variation of the measurand occurs at t = T_0.

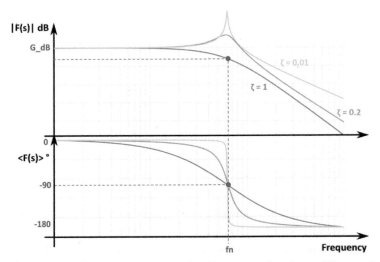

Figure 1.14   Bode diagram for a second-order sensor for three different damping coefficients.

of the frequency response tends to infinity. This behavior is shown in Figure 1.14. ζ is called damping coefficient: if it is equal to 1, the condition is called *critically damped* and the magnitude of the frequency response does not show any ripple at the natural frequency. If ζ is greater than 1, the system is called *overdamped*; and if ζ is lesser than 1, it is called *underdamped*.

Figure 1.15 shows the time domain step response for a second-order sensor for overdamped, critically damped, and underdamped conditions. If the system results underdamped, it is possible to define the *settling time*. Given an error band, which is the maximum allowable amount of residual ripple for the specific application, the settling time defines how long it takes

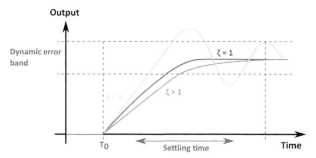

Figure 1.15 Time response of a second-order sensor output at different damping coefficients.

to the sensor to dampen the output ripples so that their amplitude remains within the error band. To conclude, it is possible to define ω_n, ζ, and S as follows:

$$\omega_n = \sqrt{\frac{k_0}{k_2}}$$

$$\zeta = \frac{1}{2}\sqrt{\frac{k_1}{k_0}} \tag{1.9}$$

$$S = \frac{1}{k_0}$$

1.3 Electronic Interfaces

As introduced in Section 1.1, not all the transducers are capable of outputting an electrical signal. It can also happen that the output voltage or current might be too low or too noisy to be adequately processed. Electronic interfaces are readout circuits introduced for sensing and manipulating (amplifying, filtering, etc.) a measurand parameter after its conversion to an electrical one. They are also capable of providing an impedance adaptation layer between the transducer and the actual load, thereby improving the overall performances of the sensor.

All the parameters that have been defined for the sensor can be used to characterize the whole sensor system, formed by a sensor and an electronic interface since that, in practice, it behaves exactly like a generic transducer. It has in fact an input parameter (for instance, resistance or capacitance variations) to be detected and quantified and an output magnitude which is typically a voltage or a current or, sometimes, a frequency. For this reason, in this section the interest will be focused on the basics of interface design showing also a number of techniques that can be employed during this stage to increase system accuracy and resolution.

1.3.1 Basic Principles

In this paragraph, for sake of simplicity, we will consider resistive sensors, but the main concepts are absolutely valid also for other types of sensors.

The simple and most basic readout technique can be implemented by means of a voltage divider (Figure 1.16). Indeed, by supposing that:

$$R_S = R_0 + \Delta R = R_0 \left(1 + \frac{\Delta R}{R_0} \right) = R_0(1 + \delta) \qquad (1.10)$$

where R_0 is the baseline (or standing value) of the sensor, it is possible to evaluate the sensor as:

$$V_{out} = V_{ref} \frac{R_0(1 + \delta)}{R_0 + R_0(1 + \delta)} = V_{ref} \frac{1 + \delta}{2 + \delta} \qquad (1.11)$$

Although very simple, this solution suffers from many drawbacks. Indeed, it has a very narrow dynamic range, being able to read not more than one decade in resistance variations; it needs to be calibrated (or in other words the sensor baseline has to be known by the designer in order to choose the value or the other resistor) and is sensitive to common mode disturbs.

To mitigate this weakness, it is possible to utilize the differential version of the voltage divider, which is the well-known bridge struc-ture (Figure 1.17(a)). Indeed, the measurand is evaluated performing the

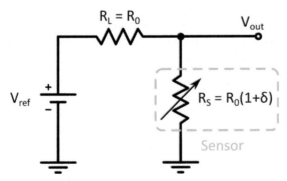

Figure 1.16   The voltage divider used as a simple interface.

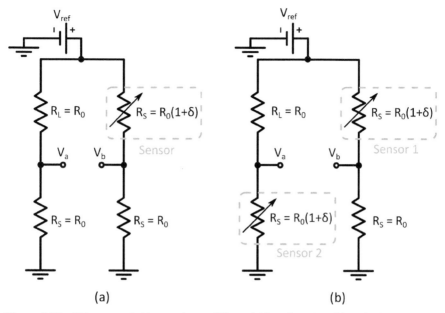

Figure 1.17 Wheatstone bridge used as a differential interface (a) with a single sensor and (b) with two sensors, one per branch.

difference between the voltages V_a and V_b inherently removing common mode error sources.

$$V_{out} = V_a - V_b = V_{ref} \left(\frac{R_0}{2R_0} - \frac{R_0}{R_0 + R_0(1 + \delta)} \right) = \frac{V_{ref}}{4} \frac{\delta}{1 + \frac{\delta}{2}}$$

$$(1.12)$$

The sensitivity $S = dV_{out}/d\delta$ of both the voltage divider and the simple bridge is constant and equal to $V_{ref}/4$ for reduced δ variations. To increase this parameter, it is possible to add a variable element also in the crossed position of the other branch of the bridge as shown in Figure 1.17(b). This allows to double the sensitivity: $S = V_{ref}/2$.

An active device (differential amplifier) can be adopted in a bridge structure to perform the difference and reconduct the signal to be of single ended kind. The bridge interface, like the voltage divider, is only suitable for small variations of the measurand, unless using more complicated structures as described in Chapter 5.

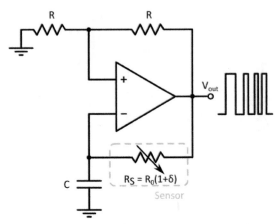

Figure 1.18   Wide variations interface implemented by an oscillator configuration.

A simple way to develop a wide input range interface is the use of a waveform generator, where the sensing resistor is capable of changing its output frequency (Figure 1.18). Supposing that the Op-Amp shows an ideal behavior, the measurand can be evaluated as:

$$T_{out} = 2.2\, R_0(1+\delta)C \tag{1.13}$$

Although the output frequency is limited by the Op-Amp bandwidth, this configuration is advantageous when the baseline of the sensor is unknown (i.e., this kind of interfaces can also work if uncalibrated) and when the signal has to be processed digitally, since the sensor value can be acquired without the need of analog-to-digital converters (ADCs).

Concerning the kind of active devices to be used in interfaces, there are several possibilities according to the particular application. The operational amplifier (Op-Amp) is probably the most versatile active block when discrete off-the-shelf-component interfaces have to be designed. Indeed, it has a high open loop gain, and according to the needed parameters, there are low-voltage rail-to-rail, low offset, low noise Op-Amps. Bandwidth can be a limiting factor only in some cases. From an integrated point of view, transconductance amplifiers (OTAs) often replace Op-Amps. Current and voltage conveyors (CCIIs and VCIIs, see also Appendix) are typically used in current mode applications. When we design interfaces that make use of active devices, we have to consider the readout error deriving from their nonidealities such as finite gain, offset, and slew rate limitations. A number of techniques, however, can be used to solve these drawbacks.

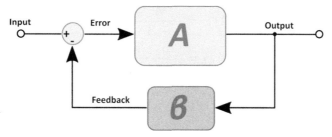

Figure 1.19 The typical block scheme for a feedback system.

1.3.2 The Feedback in Sensor Interface

One of the techniques that can be employed to maximize the accuracy of an interface is the feedback. In fact, the finite gain of an active device is one of the main sources of uncertainties during the measurements. The use of a feedback network represents a simple solution to overcome these problems. The typical block diagram of a feedback network is shown in Figure 1.19 where block A represents a high and variable (for instance, owing to environment conditions) gain path (made of at least one active device) and the β block is often made of accurate and stable passive components. The transfer function for the feedback system is then calculated as:

$$A_f = \frac{A}{1 + A\beta} \tag{1.14}$$

It is clear that if the product $A\beta$ is much greater than unity, it is possible to effectively approximate Equation (1.14) as follows:

$$A_f \cong \frac{1}{\beta} \tag{1.15}$$

Equation (1.15) shows that, for high values of A, even if it is a greatly variable quantity (for whatever reason), the closed loop gain is only fixed by the β network. It is important to underline that, even though this approximation stands, the accuracy of the closed loop system is still influenced by the finite gain of stage A *(gain error)*. In fact, by calling the ideal transfer function $A_{f,I}$ (infinite gain A) and the real one $A_{f,C}$ (high but finite gain A):

$$A_{f,I} = \frac{A}{1 + A\beta} \cong \frac{1}{\beta}$$

$$A_{f,C} = \frac{A}{1 + A\beta} \tag{1.16}$$

$$A_{f,C} - A_{f,I} = -\frac{1}{(1 + A\beta)\beta}$$

Then, by supposing that $A\beta \gg 1$:

$$A_{f,C} - A_{f,I} = -\frac{1}{A\beta^2} \tag{1.17}$$

The difference $A_{f,C} - A_{f,I}$ represents the gain error (or, in general, the sensitivity error).

One of the active blocks employed to best take advantage of the feedback technique is, obviously, the operational amplifier. That is because it has a very high open loop gain, but also because, being differential, they allow to incorporate both the subtractor and A block in a single device.

Also current [9] and voltage [10] conveyors (CCII, VCII, see Appendix) exploit the feedback technique, but in a different way with respect to the Op-Amp. Indeed, they do not present any high-level feedback path because the latter is realized internally, at the transistor level. What they do to achieve a voltage or current gain is to take advantage of their dual capability of processing both voltage and current signals. By doing that, these devices are able to amplify input signals with the gain fixed by passive components [11–15].

1.3.3 Autozero Technique

The autozero technique is generally used to reduce all these nonidealities of an interface that remain almost constant during the compensation process (in particular, input offset, low-frequency noise, etc.). Indeed, this technique is one of these methods that is carried out through multiple stages or steps (i.e., it requires switches to modify the circuit topology). The working principle can be explained referring to Figure 1.20.

As visible, there are two switches which are active in mutual exclusion. During the first phase (S_1 close, S_2 open), the output signal can be evaluated as:

$$X_{outA} = [x_{in}(t_A) + X_{off} + x_{noise}(t_A)]A \tag{1.18}$$

In Equation (1.18), x_{in} is a generic voltage signal, X_{off} is a DC input offset, x_{noise} is the noise which is supposed to be constant during the entire compensation process, and A is the gain of the active block.

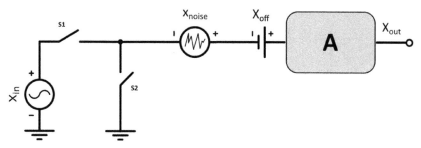

Figure 1.20 An example of autozero setup using two switches.

During the second phase (S_1 open, S_2 close), the output is given by:

$$X_{outB} = [X_{off} + x_{noise}(t_B)]A \tag{1.19}$$

By then supposing that $x_{noise}(t_B) \cong x_{noise}(t_A)$, subtracting the two signals obtained so far, it is possible to theoretically obtain an evaluation of the output signal free of offset and low-frequency noise:

$$X_{outB} - X_{outB} = Ax_{in} \tag{1.20}$$

This approach can be extended to those scenarios where more than two switches are present. In this case, it is possible to achieve higher benefits such as removing the effects of finite gain; however, the system needs to be faster in order to maintain the hypothesis formulated for the analysis. In general, as the compensation speed lowers, the effects on the noise reduction become marginal.

1.3.4 Chopper Technique

Another way to reduce the input offset as well as low-frequency noise is the so-called chopper technique. Unlike autozero circuits, this approach is realized in a time continuous operation. A chopper circuit takes its name by the presence of two *choppers* (a two-way switch, see Figure 1.21) that share the same behavior (if one is directly connected, so the other one too). The basic idea is to translate these nonidealities to a higher frequency by means of a modulation so to be effectively filtered.

The input signal on its path to the average block encounters two choppers: in particular, it is modulated by the first one and demodulated by the second one. Since offset voltage and input noise affect the input signal only after its modulation, at the output of the amplifier block, the second chopper acts as a

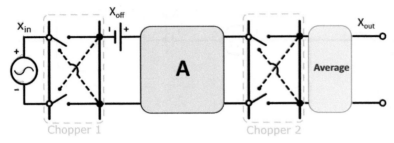

Figure 1.21   An example of chopper circuit.

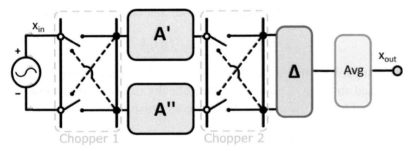

Figure 1.22   Block diagram of the dynamic element matching technique.

demodulator for the amplified input signal, whereas it acts as a modulator for both noise and offset.

With a simple filtering action (Average block), it is therefore possible to remove these two unwanted signals from the input one [2]. This technique results more effective than autozero for mitigating input noise effects. A side effect lies in its incapability of compensating the finite gain of the amplifier.

1.3.5 Dynamic Element Matching

Dynamic element matching can be used in the case when the main source of error comes from mismatches in active or passive elements that should be equal to each other. The working principle consists of dynamically interchanging mismatched components and taking the average of the output magnitude, as shown in Figure 1.22. In the example, gain of input blocks A' and A'' has slight differences. To make sure that the average gain experienced by the input signals is the same, two choppers are added in order to continuously swap the two blocks. Averaging the output of the subtractor reduces these mismatches between A' and A''.

Although similar to the chopper technique, dynamic element matching is classified as a different method. A typical use case for the dynamic element matching is to interchange input pairs in CMOS operational transconductance amplifier, where input transistors should be as similar as possible.

References

[1] J. Fraden, *Handbook of modern sensors: Physics, designs, and applications*, 5th ed. Springer, 2015.

[2] C. Falconi et al., "Electronic interfaces," *Sensors and Actuators B: Chemical*, 121(1), 295–329, 2007.

[3] D. Sheingold, *Transducer interfacing handbook*. Massachusetts: Analog Devices, Inc., 1980.

[4] I. Vikulin, V. Gorbachev and A. Nazarenko, "Radiation sensitive detector based on field-effect transistors," *Radioelectronics and Communications Systems*, 60(9), 401–404, 2017.

[5] R. Corbitt et al., *Standard handbook of environmental engineering*. New York: McGraw-Hill, 1999.

[6] J. Schroder, R. Borngraber, F. Eichelbaum and P. Hauptmann, "Advanced interface electronics and methods for QCM," *Sensors and Actuators A,* 97–98, 543–547, 2002.

[7] A. Baschirotto, E. Dallago, P. Malcovati, M. Marchesi and G. Venchi, Precise vector-2D magnetic field sensor system for electronic compass, Proceedings of IEEE Sensors, Wien, Austria, 2004, pp. 1028–1031.

[8] P. Ripka, S.O. Choi, A. Tipek, S. Kawahito and M. Ishida, "Pulse excitation of micro-fluxgate sensors," *IEEE Transactions on Magnetics*, 37, 1998–2000, 2001.

[9] G. Ferri and N.C. Guerrini, *Low voltage low power current conveyors*. Boston: Kluwer Academic Publisher, 2003.

[10] L. Safari, G. Barile, V. Stornelli and G. Ferri, "An overview on the second generation voltage conveyor: features, design and applications," *IEEE Transactions on Circuits and Systems II: Express Briefs*, 66(4), 547–551, 2019.

[11] K. Smith and A. Sedra, "The current conveyor—A new circuit building block," *Proceedings of the IEEE*, 56(8), 1368–1369, 1968.

[12] G. Ferri and N. Guerrini, *Low-voltage low-power CMOS current conveyors*. Boston, MA: Springer US, 2004.

[13] L. Grigorescu, "Amplifier built with current conveyors," *Romanian Journal of Physics*, 53, 109–113, 2008.

[14] V. Stornelli, G. Ferri, L. Pantoli, G. Barile and S. Pennisi, "A rail-to-rail constant-g_m CCII for instrumentation amplifier applications," *AEU – International Journal of Electronics and Communications*, 91, 103–109, 2018.

[15] J. A. Svoboda, "Current conveyors, operational amplifiers and nullors," *IEEE Proceedings G – Circuits, Devices and Systems*, 136(6), 317–322, 1989.

2

Capacitive Sensing

The term capacitance indicates the value of the ratio between the charge accumulated on one of at least two conductive elements close to each other in a well-determined medium and the voltage difference between these two conductive elements.

$$C = \frac{q}{V} \left[\frac{Coulomb}{Volt} \right] = [Farad] \tag{2.1}$$

The latter elements are called electrodes and the device that they form is called capacitor. According to the voltage applied to the electrodes, one of them will be charged positively (showing a lack of electrons) and the other will be charged negatively (excess of electrons). Hence, an electric field builds up between them. The shape of the field lines depends on the form of the electrodes as emphasized in Figure 2.1.

From the electrical point of view, it is possible to characterize a capacitor by using its impedance, Z_C (or admittance, Y_C), which takes into account the capacitive value and the operating frequency:

$$Z_C = R - \frac{1}{j\omega C} \cong \frac{j}{\omega C}$$

$$Y_C = G - j\omega C \cong -j\omega C \tag{2.2}$$

where R and G are the resistance and the conductance of the capacitor, respectively, j is the complex unity, and ω is proportional to the frequency f (being $\omega = 2\pi f$). The approximation in Equation (2.2) stands for capacitors with negligible parasitic resistors. Considering a capacitor as a complex impedance allows to take into consideration not only its effects on the magnitude of a signal but also its effects on its phase. Moreover, if the frequency increases, the magnitude of a capacitor becomes lower.

Looking at the capacitor from a circuital point of view, there are two possible equivalent models that trace the behavior of the device at different

25

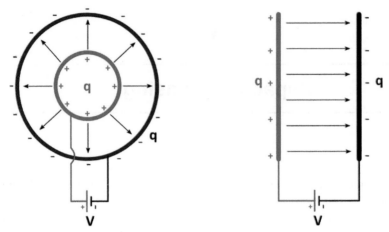

Figure 2.1 Example of capacitors with different shapes and their relative field lines.

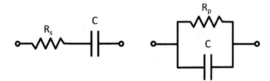

Figure 2.2 The series (left) and parallel (right) circuital equivalent model of a capacitor.

frequencies: the series model and the parallel model (see Figure 2.2). Typically, the parallel model is preferred because, unlike the series one, it is more similar to reality, being capable of modeling also the low frequency current leakage of a real capacitor.

2.1 Capacitive Sensors

2.1.1 Physical Properties

Capacitive sensors, in general, present important advantages over their resistive counterparts. Indeed, they show better sensitivity, temperature, and drift performances and, most importantly, consume virtually no power. Another noticeable advantage is that they can be integrated together with the readout electronics over a silicon substrate (MEMS capacitive sensors), making them miniaturized without losing key features like sensitivity and resolution [1–4].

An interesting aspect of a capacitor, from a sensing point of view, is that its value is influenced by its geometrical and physical properties. Taking into

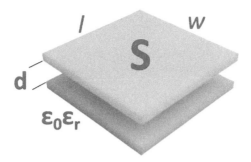

Figure 2.3 A parallel-plate capacitive sensor configuration.

consideration two parallel plates facing each other (refer to Figure 2.3), which is the most used geometry due to its easiness of fabrication, it is possible to define its capacitance as:

$$C = \varepsilon_0 \varepsilon_r \frac{S}{d} \tag{2.3}$$

where ε_0 and ε_r are the dielectric constants of the vacuum and of a specific medium, respectively; S is the overlapping surface between the two electrodes $(S = wl)$; and d is the distance between them. If an external magnitude is capable of modifying one of these parameters, a consequent variation in the capacitance value is experienced and, hence, can be read by a suitable electronic interface.

A "spacing variation capacitive sensor" (see Figure 2.4(a)) is a device where the variable magnitude is the distance between the two plates. In this case, the space variation changes the value of the same capacitance according to the following relationship:

$$C = \varepsilon_0 \varepsilon_r \frac{S}{(d + \Delta d)} \tag{2.4}$$

Since often one of the two plates is fixed, while the other is a vibrating conductive membrane, rather than the pure distance, the mean value of this magnitude across the entire plate area is given by:

$$C = \iint_{x,y} \varepsilon_0 \varepsilon_r \frac{1}{(d + \Delta d)} dx dy \tag{2.5}$$

Obviously, in Equation (2.5), the magnitude Δd is a function of x and y. This configuration results to be very sensitive to the measurand since it acts on the smallest dimension of the sensor, so a reduced variation of distance

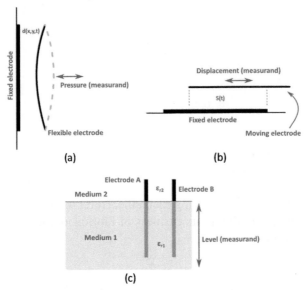

Figure 2.4   Three different use cases for a parallel plate capacitive sensor: (a) distance variations, (b) overlapping surface variations, and (c) dielectric constant variation.

can be reflected in a large variation of the capacitance. It is noticeable that variations of the capacitance with respect to the dimension d are nonlinear (Equation (2.4)); however, if the evaluated parameter is the impedance as defined in Equation (2.2), the same relationship becomes linear:

$$Z_C = \frac{d + \Delta d}{jw\varepsilon_r\varepsilon_0 S} \tag{2.6}$$

Similarly, an "overlapping area variation capacitive sensor" relies on the modification of the common surface shared by the two electrodes (see Figure 2.4(b)). An advantage of this approach with respect to the previous one is that capacitance variation is inherently linear with respect to the surface variations. On the other hand, however, since plate dimensions are much larger than the gap between them, the sensitivity of this approach results lower if compared to the previous one (in other words, to produce the same amount of capacitance change, the measurand must have a wider variation). Lastly, it is possible to take advantage of the dielectric constant of different mediums so as to achieve a capacitance variation as shown in Figure 2.4(c). This specific scenario is often used to detect liquid or gas levels into a tank, or even to reveal the presence of a particular gas in an environment. Remarkably, these techniques can be used to detect not only linearly moving electrodes but also rotation and tilt angles. More complex structures (based on arrays of parallel-plate capacitors) can act as multidimensional detectors, which means that they can simultaneously and independently sense more than one measurand variations.

2.1.2 Basic Interfaces for Capacitive Sensors

As described in Chapter 1, the interface type and performance are strictly related to the designer needs and to the application. The main key points that determine the appropriate choice for the interface are the baseline of the sensor, how broadly it can vary, how much common mode disturbs can affect the measurement (environment), how parasitic capacitances of the interface are placed with respect to the sensor, how much power can be consumed by the circuitry, and, in some cases, how much space the circuitry can occupy. An accurate analysis of the application scenario is mandatory before designing the sensor interface [4].

The first and most simple method is based on a DC readout. A basic circuit is given in Figure 2.5. The working principle is the following: the sensing capacitor is precharged to a reference voltage V_{ref}; then, by supposing that the signal varies with a frequency greater than $1/RC$, it is possible to consider the charge accumulated at the capacitor electrodes as a fixed value equal to:

$$q_{std} = C_{bl}V_{ref} \tag{2.7}$$

where C_{bl} is the value of the capacitor at the steady state (or baseline). Then, by inserting Equation (2.7) into (2.1), it is possible to evaluate the output voltage as:

$$V_{out} = \frac{C_{bl}V_{ref}}{C_x} \tag{2.8}$$

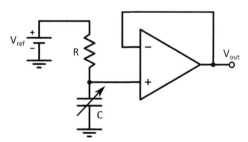

Figure 2.5 DC readout technique.

where C_x represents the value of the sensing capacitor under the action of the measurand. Although very straightforward, this method lacks robustness with respect to circuital noise, works only for relatively fast capacitor variations and requires an extremely high impedance and low offset amplifying stage. Moreover, owing to the constraint on the time constant, it is not employable for wide variations of the sensor from its baseline.

A solution that is particularly suitable for large variations from the baseline is the so-called capacitance to frequency (or period) conversion. The key point of this technique is the employment of the sensor in an oscillator configuration (even a basic RC oscillator) and the tracking of its variations by measuring changes in the output frequency. Sometimes, rather than on the frequency, output variations are induced on the duty cycle of the signal. Figure 2.6 depicts a traditional astable oscillator where C_s represents the sensing capacitor. The relationship that links the frequency to the capacitor is the following:

$$f_{out} = \frac{1}{2R_3 C_s \ln\left(1 + 2\frac{R_1}{R_2}\right)} \tag{2.9}$$

According to the oscillator nature (RC, LC, and so forth), it is possible to obtain different frequency ranges. As noticeable from Equation (2.9), RC oscillators produce an output frequency that is proportional to 1/RC, whereas for LC oscillators, the output frequency is proportional to $1/\sqrt{LC}$, so the designer has to take into account the output nonlinearity.

This kind of readout technique has numerous advantages. One of the most noticeable advantages is that they do not require knowing the capacitor

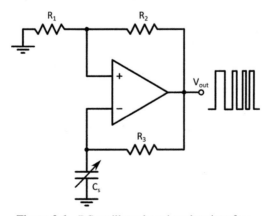

Figure 2.6   RC oscillator-based readout interface.

baseline to perform the readout, and this reflects in the possibility to achieve good measurements even if the sensor drifts with time.

Another benefit is that the output is intrinsically digital: measuring a time, for a microcontroller or a single board computer, does not require the presence of ADCs and hence reduces the system complexity and the conversion error.

On the other hand, drawbacks of oscillator-based interfaces are their sensitivity to stray capacitances and electronic mismatches and their poor noise rejection. This can lead to accuracy errors and, in general, requires specific precaution at design stage, like inserting shielding lines and complicating the actual interface. Another critical aspect is that an oscillator-based interface has typically a low sensitivity value, which makes it unsuitable for extremely low capacitance variations from the baseline value.

One of the best methods to detect very low measurand variations, maintaining high-accuracy levels, is to make use of synchronous demodulation architectures. The basic idea here is to excite the sensor by a high-frequency signal (typically a sinusoidal signal with a frequency greater than 10 kHz) and then, through a suitable circuitry, to bring back the measurement output to the baseband through a demodulation stage. This process, although intricate at a first glance, shows many benefits. First, exciting a capacitor with a sufficiently high frequency reduces its impedance. This allows to relax the high impedance requirements for the primary conversion stage, minimizing matching errors. Second, the presence of a low-pass filter during the demodulation process allows to cut down higher frequencies noise. Third, being an AC excited technique, part of the parasitic capacitances is inherently nullified (this will be discussed deeper in the following sections and chapters).

A simple circuit that exploits the synchronous demodulation is given in Figure 2.7. The sensor is coupled with a reference capacitor through a voltage divider. A simple non-inverting buffer reads the voltage at the output of the divider and feeds it to a demodulator. This consists of a multiplier and a filter: the multiplier uses a reference signal with the same frequency and phase respect to the buffer output (synchronous) to generate the DC output together with an unwanted high-frequency signal, which is removed by the filter. An improved version of the same interface is shown in Figure 2.8. Rather than relying on a simple voltage divider, it uses a bridge-based architecture to perform a fully differential measurement in order to improve the rejection to common mode disturbs, thus increasing the resolution of the whole interface.

The best choice for the first stage is, in this case, an instrumentation amplifier (INA) since, unlike a simple Op-Amp, it offers a suitably high

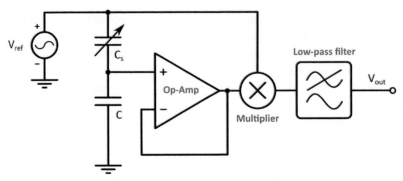

Figure 2.7   Single-ended synchronous demodulation technique applied to capacitive sensing.

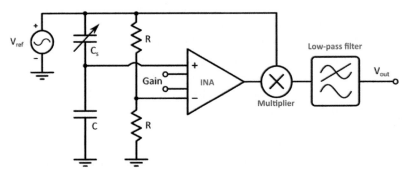

Figure 2.8   Fully differential synchronous demodulation technique.

impedance at both inputs while converting the signal to single-ended. The voltage amplification is fixed through an external resistance to be connected between the two "Gain" terminals.

There are several techniques to use a bridge-based architecture also for wide varying capacitive sensors, which will be analyzed in the next chapters.

2.2 Differential Capacitive Sensors

A differential capacitive sensor, schematized as in Figure 2.9, is an important subset of the capacitive sensors family. It is a three-terminal device made of a couple of capacitors, C_1 and C_2, which share a common node (called D). The value of these two capacitors varies from a common baseline in a differential fashion under the action of the measurand. The main advantage of this behavior is the elimination of common mode disturbs from the readout process, thus increasing the overall sensitivity and resolution of the sensor.

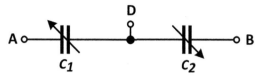

Figure 2.9 Ideal equivalent model for a differential capacitance sensor.

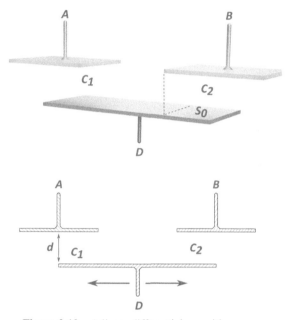

Figure 2.10 A linear differential capacitive sensor.

This makes them suitable in applications where low capacitive variations are induced by the measurand for the determination of acceleration, dielectric characteristics of a medium, angular velocity, displacement, and so on [5–7].

2.2.1 Physical Properties

According to the physical behavior of differential capacitive sensors in response to the measurand, it is possible to divide them into two main categories: linear sensors and hyperbolic sensors.

Figure 2.10 represents a linear differential capacitive sensor: it consists of three plates, two of them are fixed (the upper ones), while the third one (the bottom one) is free to move. When no external magnitude is applied, they lay

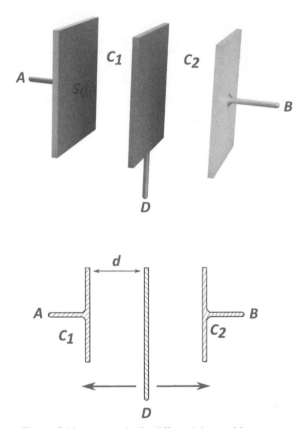

Figure 2.11   A hyperbolic differential capacitive sensor.

at a certain distance d, sharing a certain amount of area S_0. These conditions define the baseline of the sensor, C_{bl}, which can be calculated according to Equation (2.3). The action of the measurand here determines a differential change in the overlapping areas of the three plates.

Unlike the simple capacitive sensors, it is possible to uniquely define a range of variations for the sensor:

$$0 < C_{1,2} < 2C_{bl} \tag{2.10}$$

Figure 2.11 represents a hyperbolic differential capacitive sensor. The baseline condition can be calculated with the same procedure highlighted for the linear sensors. However, for this kind of sensors, the measurand acts on the distance between the three plates: the external ones are typically fixed, while the internal one is movable. As a consequence, the range where a hyperbolic sensor varies can be estimated as:

$$\frac{C_{bl}}{2} < C_{1,2} < \infty \tag{2.11}$$

Based on what has been analyzed so far, it is possible to make some considerations. As already pointed out in the previous paragraphs, a hyperbolic sensor tends to be more sensitive to the measurand respect to the linear counterpart due to the fact that, for any given displacement, the smallest dimension (d) is much more affected in terms of percentage variation than the largest one. This makes a linear sensor more suitable for measurands with large variations.

From Equations (2.10) and (2.11), the difference in the ranges that the two different options can assume is noticeable. On the one hand, linear sensors are easier to interface with, due to the finite variation span. On the other hand, however, the fact that one of the two capacitors can assume very low values makes them more dependent on the effects of parasitic capacitors in some applications. For both the options, the extreme boundary conditions cannot be reached because the differential capacitor would degenerate into a simple capacitor.

2.2.2 Parametrization

Regardless of the type of sensor, due to the intrinsic differential behavior, the evaluation of the measurand can be pursued by evaluating a dimensionless parameter x rather than the actual capacitance variation. This parameter can be defined as the ratio between the difference of the two capacitors C_1 and C_2 of the sensor, and their sum:

$$x = \frac{C_1 - C_2}{C_1 + C_2} \tag{2.12}$$

In other words, x defines how much the sensor value has changed relatively to the baseline, and therefore, it is often expressed as a percentage. As easily noticeable, its variations are bounded between -1 and 1, values that, as visible from Equations (2.10) and (2.11), can be never reached.

From an analytical perspective, it is therefore advantageous to parametrize the behavior of a differential capacitance sensor with respect to x. Equation (2.13) shows how a linear sensor can be expressed as a function of x:

$$C_1 = C_{bl}(1 + x)$$
$$C_2 = C_{bl}(1 - x) \tag{2.13}$$

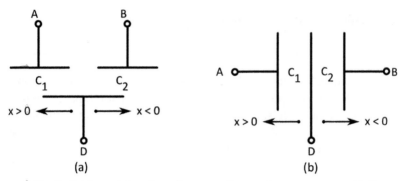

Figure 2.12   Definition of the sign of x according to the sensor type: (a) linear and (b) hyperbolic.

Positive or negative values are related to the representation of Figure 2.12(a).

Similarly, Equation (2.14) shows how a hyperbolic sensor can be parametrized:

$$C_1 = C_{bl}\frac{1}{1-x}$$

$$(2.14)$$

$$C_2 = C_{bl}\frac{1}{1+x}$$

The positive and negative signs are considered according to Figure 2.12(b).

2.2.3 Basic Interfaces for Differential Capacitive Sensors

The basic solutions that have been reported in Section 2.1.2 for capacitive sensors can be applied to differential capacitance sensors as well. Figure 2.13 represents two simple oscillators implementing a capacitance to time conversion. The linear capacitive sensor can be used in the interface of Figure 2.13(a), while the hyperbolic one can be used in the interface of Figure 2.13(b). This is due to the intrinsic behavior of the sensor: for the linear one, the parallel configuration would remain constant across all the measurand variations making the interface unsensitive to the measurand.

Analogously, for the hyperbolic sensor, the series configuration would result in a constant output frequency regardless of the measurand.

This is a very basic solution. Indeed, it is unable to detect which of the two capacitors has a greater value. In other words, it does not take advantage of the differential feature of the sensor.

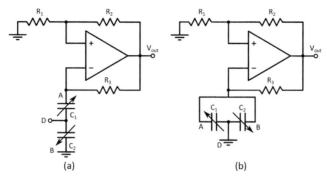

Figure 2.13 Oscillator-based differential capacitive sensor interface (a) for linear sensors and (b) for hyperbolic sensors.

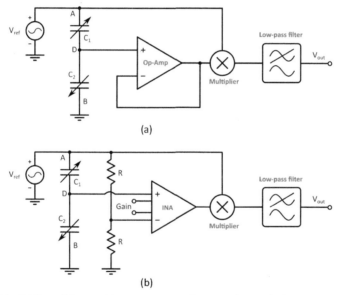

Figure 2.14 Differential capacitive sensors asynchronous demodulation readout: (a) single-ended and (b) fully differential.

On the contrary, synchronous demodulation configurations can benefit of a differential capacitance sensor. They are shown in Figure 2.14(a) and (b) where we have reported both the single-ended and the fully differential amplifier configurations, respectively. The presence of a differential capacitive sensor allows to obtain a better sensitivity and an improved resolution (due to its lower noise).

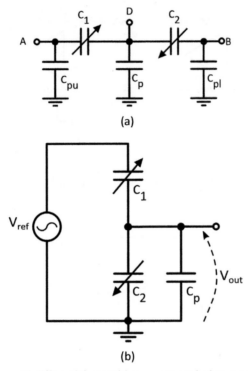

Figure 2.15   Real-world differential capacitive sensor equivalent model: (a) generic and (b) driven by an AC source.

2.2.4 Consideration on Parasitic Capacitances

One of the key points that, with the scaling of the technology, has become a necessity in interfaces suitable for very low baseline sensors (as, for example, MEMS) is their capability to deal with sensor parasitic impedances [8]. In fact, there are cases where stray capacitances are even greater than the actual sensor baseline.

Figure 2.15(a) shows the equivalent model of a real-world differential capacitive sensor. As visible, there are three main contributions to the overall stray impedances: C_{pu}, C_p, and C_{pl} [9, 10].

The mitigation of the effects of both C_{pu} and C_{pl} is typically exploited by means of a suitable topology: for instance, the use of an AC-driven setup inherently mitigates both of them since (see Figure 2.15(b)) C_{pl} is grounded, while C_{pu} is connected to a very low impedance signal generator. The following considerations will then be focused on the effects of C_p only.

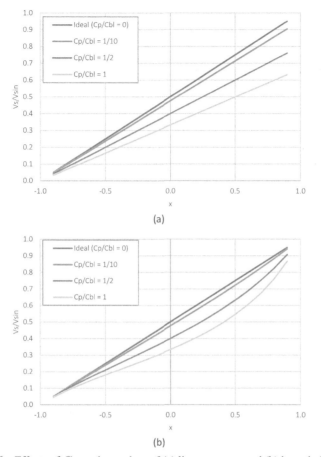

(a)

(b)

Figure 2.16 Effects of C_p on the readout of (a) linear sensors and (b) hyperbolic sensors.

From this figure, we can write that:

$$V_{out} = V_{ref} \frac{C_1}{C_1 + C_2 + C_p} \qquad (2.15)$$

Therefore, by substituting Equations (2.13) and (2.14) in (2.15), we can find the relationship between the input voltage (V_{ref}) and the output voltage (V_{out}) for linear sensors as:

$$V_{out} = V_{ref} \frac{1 + x}{2 + \frac{C_p}{C_{bl}}} \qquad (2.16)$$

and for hyperbolic ones as:

$$V_{out} = V_{ref} \frac{\frac{1}{1-x}}{\frac{2}{1-x^2} + \frac{C_p}{C_{bl}}} \tag{2.17}$$

Ideally ($C_p = 0$), both equations tend to the same linear one. However, under the presence of parasitic capacitances, the readout circuit suffers a decrease in sensitivity (maintaining a linear behavior) in the case of linear sensors (see Figure 2.16(a)), while in the case of hyperbolic ones, linearity worsens if the ratio C_p/C_{bl} increases (see Figure 2.16(b)). This analysis holds for more complex interfaces.

Advanced interfacing techniques employing a mitigation of the effects of parasitic capacitances will be discussed in the following chapters.

References

[1] J. Fraden, *Handbook of modern sensors: Physics, designs, and applications*, 5th ed. Springer, 2015.

[2] L. K. Baxter, "Capacitive sensor basics," in *Capacitive sensors: Design and applications*. New York, NY, USA: The Institute of Electrical and Electronics Engineers, pp. 1–46, 1997.

[3] B. Granados-Rojas, M. A. Reyes-Barranca, G. S. Abarca-Jiménez, L. M. Flores-Nava and J. A. Moreno-Cadenas, "3-layered capacitive structure design for MEMS inertial sensing," 2016 13th International Conference on Electrical Engineering, Computing Science and Automatic Control (CCE), Mexico City, 2016, pp. 1–5.

[4] R. Puers, "Capacitive sensors: When and how to use them," *Sensors and Actuators A: Physical*, 37–38, 93–105, 1993.

[5] H. Tavakoli, H. G. Momen and E. A. Sani, "Designing a new high performance 3-axis MEMS capacitive accelerometer," 2017 Iranian Conference on Electrical Engineering (ICEE), Tehran, 2017, pp. 519–522.

[6] T. Kose, Y. Terzioglu, K. Azgin and T. Akin, "A single-mass self-resonating closed-loop capacitive MEMS accelerometer," 2016 IEEE SENSORS, Orlando, FL, 2016, pp. 1–3.

[7] Y. Terzioglu, T. Kose, K. Azgin and T. Akin, "A simple out-of-plane capacitive MEMS accelerometer utilizing lateral and vertical electrodes for differential sensing," 2015 IEEE SENSORS, Busan, 2015, pp. 1–3.

[8] S. Baglio, S. Castorina, G. Ganci and N. Savalli, "A high sensitivity conditioning circuit for capacitive sensors including stray effects

compensation and dummy sensors approach," Proceedings of the 21st IEEE Instrumentation and Measurement Technology Conference (IEEE Cat. No. 04CH37510), Como, 2004, pp. 1542–1545, Vol.2.

[9] N. A. C. Mustapha, A. H. M. Z. Alam, S. Khan and A. W. Azman, "Single Supply Differential Capacitive Sensor with Parasitic Capacitance and Resistance Consideration," 2018 7th International Conference on Computer and Communication Engineering (ICCCE), Kuala Lumpur, 2018, pp. 445–448.

[10] J. Shiah, H. Rashtian and S. Mirabbasi, "A low-noise parasitic-insensitive switched-capacitor CMOS interface circuit for MEMS capacitive sensors," 2011 IEEE 9th International New Circuits and systems conference, Bordeaux, 2011, pp. 470–473.

3

Voltage Mode Differential Capacitive Sensor Interfaces

In this chapter, some approaches related to the design of interface electronic circuits of a differential capacitive sensor will be analyzed, highlighting benefits as well as drawbacks that have to be carefully considered during the design stage. The analysis will be focused on the voltage mode interfaces. Scientific society has often debated on the actual meaning of voltage mode versus current mode [1, 2]; however, in this chapter we will consider as voltage mode all these techniques where the majority of the processing of signals within the actual interface involves the manipulation of voltages, or in other words, all these interfaces where, even if there is the presence of a current signal, this is converted into a proportional voltage for further processing. This is why, in this chapter, circuits employing charge sensitive amplifiers (CSAs) will also be analyzed.

The discussion will be addressed considering the following four macro-approaches: the *step measurement*, where the sensing process involves many preparatory phases prior to the actual readout; the *oscillator-based* capacitance to time conversion; the *time continuous* capacitance to voltage conversion; and, finally, the *capacitance to digital* conversion. The output of the latter three groups is well defined (a time, a voltage, or a binary code), whereas the first group can have either a time or a voltage output.

Both discrete and integrated solutions will be analyzed, but only focusing on their working principle.

3.1 Step Measurement

As the name suggests, these techniques involve the use of switches and driving clocks through which it is possible to modify the topology of the

43

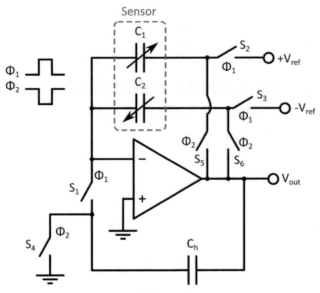

Figure 3.1   Sample and hold-based interface [3].

circuit to adequately carry out one of at least two steps to perform the differential capacitance readout.

The first proposal [3, 4] is intended to be used with micromachined transducers and hence aims to have good sensitivity, accuracy, and immunity to noise and circuit nonidealities. The equivalent schematic is shown in Figure 3.1. It can be thought of as a sample and hold where the holding capacitor is represented by the two differential sensor capacitors. There are two switching signals, Φ_1 and Φ_2, sharing the same frequency and opposite phases; therefore, the readout process involves two distinct steps.

During the first step, switches S_1, S_2, and S_3 are on, while switches S_4, S_5, and S_6 are off. Sensing capacitors are therefore precharged to V_{ref} and $-V_{ref}$, respectively. The holding capacitor C_h is added so as to provide Op-Amp gain error compensation, across the two phases together with C_1 and C_2 (this technique is thoroughly explained in Ref. [5], and it strictly resembles the correlated double sampling CDS technique). During the second step, Φ_2, switches status is inverted. This means that the two capacitors C_1 and C_2 of the sensor are connected in parallel, and therefore, neglecting nonidealities, it is possible to calculate the output voltage as follows:

$$V_{out}(\Phi_2) = \frac{Q_{TOT}}{C_{TOT}} = \frac{C_1 V_{ref} - C_2 V_{ref}}{C_1 + C_2} = x V_{ref} \qquad (3.1)$$

Taking back into consideration nonidealities, the first thing to notice is that this technique allows to neglect the effects of parasitic capacitances C_{pu} and C_{pl} which are in parallel with C_h during the measuring stage (Φ_2) because they are always connected to a low impedance node.

Due to the negative feedback, at a first glance it might appear that C_p could also be neglected; however, the effects of a finite gain might make this statement wrong. As a consequence, the effects to be taken into consideration are Op-Amp finite gain A and offset, as well as charge injection which is a typical drawback of switched capacitors topologies. Considering the n-th cycle, the output voltage for Φ_1 and Φ_2 can be computed as:

$$V_{out}\left(\Phi_{1,n}\right) = V_{out}\left(\Phi_{2,n-1}\right)$$
$$+ \frac{1}{C_h} \frac{\left(C_1 + C_2\right) V_{out}\left(\Phi_{2,n-1}\right) - \left(C_1 - C_2\right) V_{ref}}{1 + \frac{C_h + C_1 + C_2 + C_p}{A C_h}} \quad (3.2)$$

$$V_{out}\left(\Phi_{2,n}\right) = \frac{\left(C_1 - C_2\right) V_{ref} + \frac{\left(C_1 + C_2 + C_p\right) V_{out}\left(\Phi_{1,n}\right)}{A}}{\left(C_1 + C_2\right)\left(1 + \frac{C_h + C_1 + C_2 + C_p}{A C_h}\right)} \quad (3.3)$$

Under the condition $C_h \gg C_1 + C_2$, Equation (3.2) can be simplified as follows:

$$V_{out}\left(\Phi_{1,n}\right) = V_{out}\left(\Phi_{2,n-1}\right) \quad (3.4)$$

Substituting Equation (3.4) in (3.3), it is possible to simplify it as follows:

$$V_{out}\left(\Phi_{2,n}\right) = \frac{\left(C_1 - C_2\right) V_{ref}}{C_1 + C_2} = x V_{ref} \quad (3.5)$$

From this analysis, it is clear that, more than a compensation, the proposed topology helps to relax the high gain condition on the active device allowing to achieve, with a lower gain, the same performances (in terms of accuracy) that are attainable only with a much higher Op-Amp gain. This is a helpful property when designing with very low pitch technologies (where high gain is hard to be achieved). On the other hand, the main limitation of the interface is due to the charge injection of switches S_1, S_2, and S_3 on the sensor capacitors. This can be estimated as a "parasitic" voltage, given by:

$$\delta V_{out}(\Phi_2) = \frac{Q_{inj}}{C_1 + C_2} \quad (3.6)$$

The amount of total injected charges Q_{inj} strictly depends on the adopted technology parameters and can be mitigated by decreasing the dimensions of

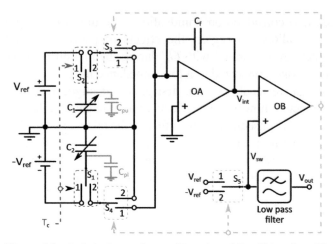

Figure 3.2   Switched capacitor oscillator-based C to V interface [6].

the switches. Since this interface has an extremely simple design, it can be easily integrated.

Another possible approach concerns the use of an oscillator and relies on its switching frequency to execute the ratiometric operation that defines the x parameter [6]. This is the working principle of the following interface which, anyway, performs a capacitance to voltage conversion. This means that the magnitude that is proportional to the x parameter remains a voltage amplitude.

The circuit solution is shown in Figure 3.2 and consists of a switched inverting charge sensitive amplifier (OA), a comparator (OB), and a low-pass filter. The CSA, together with the comparator, implements a relaxation oscillator. The low-pass filter is responsible for the generation of output voltage. In the circuit, there are 5 switches: S_1 and S_2 are driven by a fixed external clock signal with a period T_c and 50% duty cycle allowing the sensor capacitors to charge and discharge at a fixed rate, while S_3, S_4, and S_5 are driven by the oscillator itself. The two reference voltages V_{ref} and $-V_{ref}$ are used to precharge C_1 and C_2 and to generate the output signal. S_3 and S_4 allow that only one between C_1 and C_2 is connected at the OA amplifier at a time.

The interface works as follows: let us suppose as a starting condition that the output of the integrator is positive, as well as its first-order derivative. Switches S_3, S_4, and S_5 are in the 1 position, while S_1 and S_2 change between 1 and 2 at each clock cycle (1 at the rising edge, 2 at the falling edge). As long as they remain in position 1, C_1 is charged to $+V_{ref}$, while C_2 is charged to $-V_{ref}$. When the switches toggle to position 2, C_1 is grounded, while C_2 is connected to the CSA. This means that at each clock cycle, the feedback capacitor C_f receives part of the charges stored in C_2, and therefore, the output voltage of the integrator increases by a fixed step:

$$V_{int}(n) = V_{int}(n-1) + V_{step+}$$

$$V_{step+} = V_{ref}\frac{C_2}{C_f} \tag{3.7}$$

This trend continues as long as the integrator output reaches the threshold of the comparator V_{ref} when S_5 is on the position 1. Once this condition is reached, the comparator output V_c becomes low and switches S_3, S_4, and S_5 are driven to position 2. In this condition at each external clock cycle, capacitor C_1 is connected to the integrator, which, as a consequence, starts decreasing its output:

$$V_{int}(m) = V_{int}(n) - |V_{step-}|$$

$$|V_{step-}| = |V_{ref}|\frac{C_1}{C_f} \tag{3.8}$$

This situation, again, stands as long as the output of OA reaches the comparator's negative threshold. The behavior described so far repeats until unperturbed. The time T_H that V_{int} takes to ramp up to V_{ref} depends therefore on C_2, while the time T_L that it takes to ramp down to $-V_{ref}$ depends on C_1, as shown in Figure 3.3. From Equations (3.7) and (3.8), it is possible to

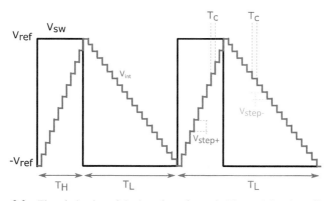

Figure 3.3 Time behavior of the interface shown in Figure 3.2, when $C_1 > C_2$.

calculate T_H and T_L as:

$$T_H = T_c \frac{2V_{ref}}{V_{step+}} = T_c \frac{2C_f}{C_2} \tag{3.9}$$

$$T_L = T_c \frac{2V_{ref}}{V_{step-}} = T_c \frac{2C_f}{C_1} \tag{3.10}$$

Taking into consideration the low-pass filter input voltage V_{sw}, this signal oscillates between V_{ref} and $-V_{ref}$ according to the T_H and T_L times (see Figure 3.3) so, by setting a sufficiently low cutoff frequency $fc \ll 1/(T_H + T_L)$, it is possible to express its output as:

$$V_{out} = \frac{1}{T_H + T_L} \left(\int_0^{T_H} V_{ref} dt + \int_{T_H}^{T_H+T_L} -V_{ref} dt \right)$$

$$= \frac{T_H - T_L}{T_H + T_L} V_{ref} = \frac{C_1 - C_2}{C_1 + C_2} V_{ref} = x V_{ref} \tag{3.11}$$

Referring to Figure 3.2, C_f should be made much greater than the sensor baseline. This allows to both increase the overall resolution of the interface and reduce the overshoot and undershoot errors (it happens when the ratio $2C_f/C_{1,2}$ is not an integer). A large C_f, however, also increases readout times. A good trade-off can be $C_f/C_{bl} = 200$. Switches' resistances are negligible since the variation induced in the charging and discharging time constants is negligible as well. Unlike the previous solution, in this interface the effects of stray capacitances have to be taken into account. Particularly, C_p can be neglected since it is always grounded. C_{pu} and C_{pl}, however, affect the measurement since they modify T_h and T_l introducing both a loss of sensitivity and an offset:

$$V_{out} = \left(\frac{C_1 - C_2}{C_1 + C_2 + C_{pu}} + \frac{C_{pu} - C_{pl}}{C_1 + C_2 + C_{pu}} \right) V_{ref} \tag{3.12}$$

Concerning charge injection and clock feedthrough, the same considerations as before are applicable to this scenario. Remarkably, by taking V_{sw} as output voltage and performing the ratio of times digitally, this interface suits well as a first stage of a digital system since it would not require ADCs to extract the information.

A similar approach that is expressively developed to deal with parasitic capacitance compensation is shown in Figure 3.4 [7].

The interface consists of a charge amplifier and eight switches that work in different steps, driven by two clocks sharing the same frequency $(1/T)$

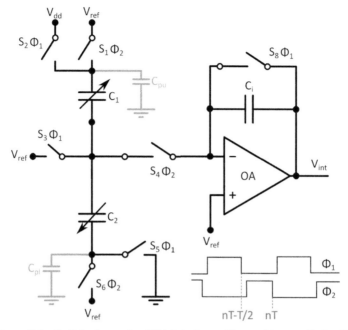

Figure 3.4 Switched capacitor CSA interface with parasitic cancellation [7].

and opposite phase. During the first stage (Φ_1 high), switches S_2, S_3, S_5, S_6, and S_8 are "ON". Capacitor C_i is therefore discharged, while it is possible to calculate the total amount of charges stored in this configuration (using V_{ref} as reference potential), as follows:

$$Q_{tot\Phi_1} = \frac{V_{DD}}{2}C_1 + \left(-\frac{V_{DD}}{2}C_2\right) + V_{DD}C_{pu} \qquad (3.13)$$

Noticeably, C_p and C_{pl} are not present into Equation (3.13): indeed, C_{pl} is grounded, so it does not contribute to the overall charge, while C_p, although charged to V_{ref}, does not experience boundary condition changes during Φ_2 and therefore is not present in Equation (3.13). At time equal to $nT - T/2$, Φ_2 goes high, while Φ_1 turns low.

It is possible to compute the total charge stored in this new configuration as:

$$Q_{tot\Phi_2} = \frac{V_{DD}}{2}(C_T + C_{pu} + C_{pl} + (1 + A)C_i) - (C_T + (1 + A)C_i)V_i$$

$$(3.14)$$

where C_t is given by $C_1 + C_2$ and A is the amplifier gain. Now, considering that the output voltage at any time can be calculated as:

$$V_{int} = A \left(\frac{V_{DD}}{2} - V_i \right) \tag{3.15}$$

and that the total amount of charges in both phases has to remain equal, it is possible to evaluate the output of the interface as:

$$V_{int}(nT) = \frac{AV_{DD}(C_{pu} - C_{pl} - C_i + 2\Delta C)}{2(C_T + C_i(1 + A))} \tag{3.16}$$

In Equation (3.16), the x parameter is contained in the ΔC factor at the numerator, and as visible, unlike (3.12), parasitic capacitors are absent from the denominator. Therefore, the sensitivity error is inherently removed from the measurement. The offset is highly mitigated from the fact that, due to fabrication processes, often the values of C_{pu} and C_{pl} are very similar, and hence, they compensate themselves.

The same configuration can be further enhanced at the expense of circuit complexity. This architecture, reported in Figure 3.5, is able to cancel out the offset error from the output voltage [7, 8].

This version basically provides a modulated output which is achieved by switching upper and lower nodes (where C_{pu} and C_{pl} are located) from V_{dd} to GND and vice versa, thanks to the addition of new switches and clock signals. By doing this, the information obtained from the difference between C_1 and C_2 is stored in the pass-band (rather than in the baseband, as it was in the previous interface). Once the signal is demodulated back to the baseband, the offset results, therefore, are ideally removed. Also $1/f$ noise is removed since, similarly to the offset, it is filtered out during the demodulation process. The peak-to-peak value of the output is equal to:

$$V_{int,pp} = \frac{2V_{DD}\Delta C}{C_i} \tag{3.17}$$

Techniques based on a charge sensitive amplifier can be used in a differential fashion as in Refs. [9, 10]. Figure 3.6 depicts such a scenario, where a fully differential switched capacitor CSA-based interface is shown.

Analogously to the first proposed interface, the latter one makes use of the correlated double sampling technique to reduce amplifier gain error and offset. The interface consists of a bridge-like structure made of sensor capacitors C_1 and C_2 and reference capacitors C_r. C_f is the feedback capacitor of

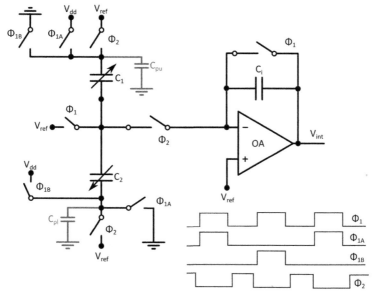

Figure 3.5 Switched capacitor interface with parasitic cancellation and modulation enhancement.

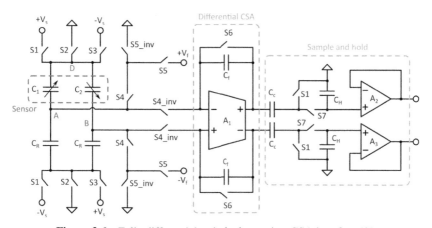

Figure 3.6 Fully differential switched capacitor CSA interface [9].

the CSA, while C_c and C_h work with the sample and hold structures in order to provide an easy analog-to-digital conversion of the output.

According to the switches status shown in Figure 3.7, the circuit executes a readout in four steps: *discharge* (or *reset*), *sensing A (Charge)*, *sensing B*

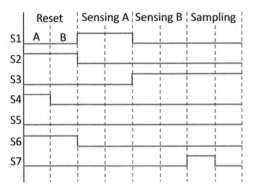

Figure 3.7   Switches status of the analyzed interface according to the relative phase of the readout.

(*Charge*), and *sampling*. The two sensing phases are used to carry out the CDS algorithm, so the errors coming from Op-Amp nonidealities are stored and then compensated.

The working principle of the interface is the following. During the *reset* phase, S_2 and S_6 are on, while S_4 toggles between on and off. This ensures the full discharge of sensor capacitors, reference capacitors, and feedback capacitors. Charge *sensing* 1 phase starts with switches S_1, S_2, and S_6 toggling their state.

This makes sure that the bridge is biased between the two reference voltages $+V_s$ and $-V_s$. Under these circumstances, the differential output of the CSA is given by:

$$\Delta V_{out,1} = V_{error} - V_s \frac{C_1 - C_2}{C_f} \tag{3.18}$$

During the *second sensing* stage, the switches S_1 and S_3 are toggled, and therefore, the reference voltages are swapped. The CSA output is given by:

$$\Delta V_{out,2} = V_{error} + V_s \frac{C_1 - C_2}{C_f} \tag{3.19}$$

Finally, during the *sampling phase*, switch S_7 is activated. The differential output voltage of the sample and hold results proportional to $C_1 - C_2$ and independent from the voltage V_{error}:

$$\Delta V_{out,SH} = \Delta V_{out,1} - \Delta V_{out,2} = 2V_s \frac{C_1 - C_2}{C_f} \tag{3.20}$$

Due to the differential structure, the interface is less sensitive to common mode disturbs; however, switches have to be carefully designed and driven in order to minimize charge injection and clock feedthrough.

3.2 Switchless Capacitance to Time Converters

In this section, capacitance to time interfaces will be analyzed. Unlike in the previous one, they do not employ clock-driven switches.

The first circuit, presented in Ref. [11], is shown in Figure 3.8. It consists of a voltage integrator (*A1*), characterized by a variable feedback path, an inverting voltage adder (*A2*), and two comparators (*A3* and *A4*). The integrator together with the variable threshold comparator (*A3*) performs as a relaxation oscillator, while *A4* acts as a simple voltage squarer. For the following analysis, diodes D_1 and D_2, which are in the negative feedback path of an amplifier, can be considered ideal. Since they are back-to-back positioned, only one of them at a time can be active, according to voltages V_{c1}, V_{c2}, and V_1. Resistances R_{1-3} are assumed equal to each other. As a matter of fact, we can say that *A2* is effectively a simple inverting voltage follower.

The oscillation period of the circuit is determined by the sum of four distinct times (see Figure 3.9). Let us impose as a starting condition, T_1,

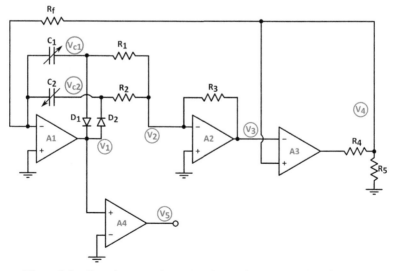

Figure 3.8   Capacitance to time relaxation oscillator-based interface [11].

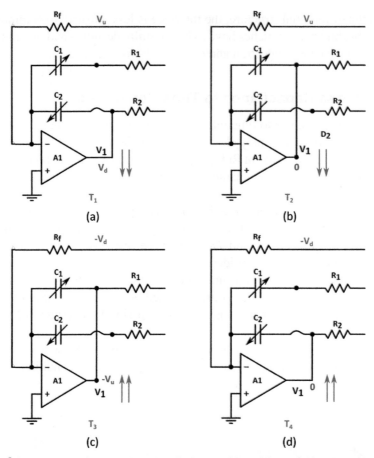

Figure 3.9   Circuit configuration based on diodes conditions: blue variables show the starting condition for V1, and the red arrows show the tendency of V1 during each phase. Pictures (a) to (d) represent T_1 to T_4, respectively.

that V_4 is high (V_u, referring to Figure 3.10), as well as V_1. Therefore, D_2 is active, C_2 is charged, while V_1 decreases linearly (Figure 3.9(a)). This phase lasts until V_1 reaches 0 V, and therefore, D_2 is turned off. Calling V_d the starting voltage across the capacitor C_2 (note that V_d should be equal to V_u, any imbalance is reflected in the readout process), it is possible to quantify T_1 as:

$$T_1 = C_2 R_t \frac{V_d}{V_u} \tag{3.21}$$

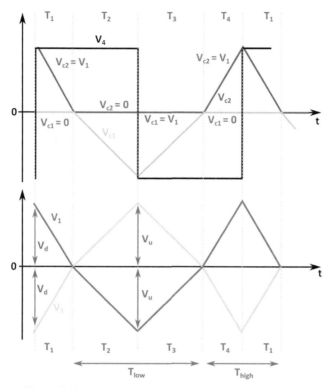

Figure 3.10 Time domain behavior of the interface signals.

During T_2 phase (Figure 3.9(b)), diode D_1 is active. Since V_3 is still lower than V_4, the latter voltage remains high, equal to V_u, and therefore, the output of the integrator continues decreasing with a slope that depends on C_1. Since this phase is not influenced by the voltage because the integrator is driven by the threshold voltage of the comparator $A3$, the time that is taken to V_1 to reach this threshold voltage $(-V_u)$ is given by:

$$T_2 = C_1 R_t \tag{3.22}$$

At this stage, V_4 changes polarity and the same for the threshold voltage of the comparator $A3$ (goes to $+V_u$). This means that the voltage V_1 now starts raising back (Figure 3.9(c)). These are the starting conditions for T_3, where D_1 is still conducting. This phase concludes when V_1 reaches 0 V and the two diodes toggle their states. Time T_3 is equal to:

$$T_3 = C_1 R_t \frac{V_u}{V_d} \tag{3.23}$$

The last phase, T_4 (Figure 3.9(d)), returns all the signals back to the starting conditions: V_4 is negative and so the output of *A1* keeps on raising until V_3 reaches the threshold voltage of *A3*. This phase lasts for a time given by:

$$T_4 = C_2 R_t \qquad (3.24)$$

From here, this cycle keeps on repeating indefinitely. To evaluate the measurand, it is possible to measure the duty cycle (D) of V_5:

$$D = \frac{T_{high}}{T_{high} + T_{low}} = \frac{T_1 + T_4}{T_1 + T_2 + T_3 + T_4}$$

$$= \frac{C_2\left(\frac{V_d}{V_u} + 1\right)}{(C_1 + C_2)\left(1 + \frac{C_1}{C_1 + C_2}\frac{V_u}{V_d} + \frac{C_2}{C_1 + C_2}\frac{V_d}{V_u}\right)} \qquad (3.25)$$

As visible, the main error source for the readout is the mismatch between V_u and V_d. It is indeed possible to slightly modify the same interface in order to make it insensitive to the mismatch between upper and lower voltages. The modified interface is shown in Figure 3.11. As visible, the comparator *A4* takes V_3 and a portion of V_4 (αV_4, where $\alpha = R_6/(R_5 + R_6)$) as inputs. By then recalling that phases T_2 and T_4 were independent from the voltage, it is advantageous to define T_5 and T_6 as a portion of them, in particular when both *A3* and *A4* comparators are either high or low (see Figure 3.12):

$$T_5 = \frac{C_1 R_1 R_t (1 - \alpha)}{R_3} \qquad (3.26)$$

$$T_6 = \frac{C_2 R_2 R_t (1 - \alpha)}{R_3} \qquad (3.27)$$

It is finally possible to evaluate the ratio between the two sensor capacitors, by executing the following operation:

$$\frac{T_5}{T_5 + T_6} = \frac{C_1}{C_1 + C_2}(1 + \varepsilon) \qquad (3.28)$$

As visible, the ratio between T_5 and T_6 is independent from V_u and V_d while giving the same ratiometric evaluation.

The parameter ε this time indicates the effects of the mismatch between R_1 and R_2 and can be defined as:

$$\varepsilon = \frac{R_1 - R_2}{R_1} \qquad (3.29)$$

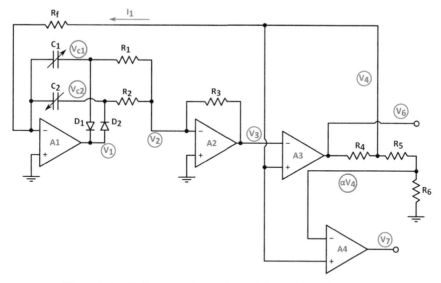

Figure 3.11 Reference voltage mismatch insensitive enhancement.

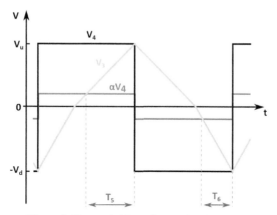

Figure 3.12 Definition of T_5 and T_6 intervals.

A similar, yet modified, approach [12] is shown in Figure 3.13. This interface still makes use of a relaxation oscillator, but instead of relying on diodes to dynamically adjust the feedback of the integrator, it uses multiple feedbacks to sustain the oscillation. Since the output stage of the interface is a mixer, it allows to incapsulate the differential capacitive readout on both its frequency and its duty cycle. The circuit is formed by three Op-Amps, which employ an integrator and two variable reference comparators. Passive

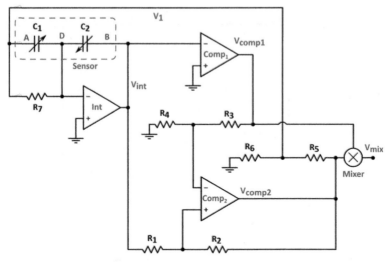

Figure 3.13   Multiple feedback C to T interface [12].

elements play a crucial role since they allow to set sensitivity, resolution, and range of the interface, other than fixing the initial oscillation of the interface together with sense capacitors.

The working principle is shown in Figure 3.14 and can be analyzed as follows. Let us impose as starting condition for phase 1 that the output of the integrator V_{int} is at V_{high} (which represents the peak value for its output). Let us also impose that the voltage that feeds the integrator, V_1, is high (i.e., V_{comp2} is equal to V_{dd}). V_1 is a square wave whose amplitude is a portion of V_{dd}. To complete the definition of the initial state, it is enough to impose that $V_{int} > V_1$ and $V_3 > V_2$, and therefore, $V_{comp1} = V_{dd}$.

All these conditions are set through the voltage dividers of the interface. From this state, the voltage V_{int} starts lowering linearly, and the same does V_3. This condition lasts as long as V_{comp1} toggles to V_{ss}. For this to happen, it is necessary that:

$$\frac{R_6}{R_5 + R_6} > \frac{R_4}{R_3 + R_4} \tag{3.30}$$

Since, however, V_{comp2} is still at V_{dd}, the voltage at the integrator output during phase 2 continues lowering. Once it reaches a level so that V_3 becomes lower than V_2, V_{comp2} toggles to V_{ss} and V_{int} starts raising again from the minimum value V_{low}.

Phase 3 starts with the raising of V_{int}. It is the same as described during phase 1: V_{comp1} and V_{comp2} remain low while V_{int} lowers, until V_{int} becomes

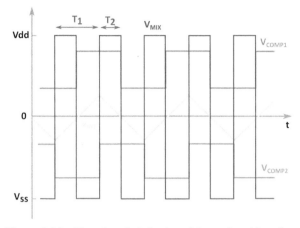

Figure 3.14 Time domain behavior of the analyzed interface.

greater than V_1 so forcing V_{comp1} to switch to V_{dd}. During the last phase of the cycle, V_{int} reaches back V_{high} closing a full period of V_{int} when V_{comp2} raises to V_{dd}.

The overall output of the interface, V_{mix}, has a frequency that is twice doubled with respect to its inputs (indeed, a multiplication produces also the sum of the input frequencies at its output); it changes according to the polarity of its inputs V_{comp1} and V_{comp2}, and therefore, during phase 1 and phase 3, it is positive, while it is negative during the other two.

Then, by evaluating the two times T_1 and T_2 representing the period and the pulse width of V_{mix}, respectively, it is possible to evaluate C_1 and C_2 as follows:

$$C_1 = \frac{1}{R_7}(T_2 - T_1)\alpha - \frac{1}{R_7}\frac{T_1}{2}$$

$$C_2 = \frac{1}{R_7}(T_2 - T_1)\beta \tag{3.31}$$

where:

$$\alpha = \frac{\dfrac{R_1}{R_1+R_2} + \dfrac{R_4}{R_3+R_4}}{\dfrac{R_6}{R_5+R_6}\dfrac{R_1}{R_2+R_1} - \dfrac{R_2}{R_2+R_1} - \dfrac{R_4}{R_4+R_3} - \dfrac{R_6}{R_5+R_6}}$$

$$\beta = \frac{\left(1 + \dfrac{R_1}{R_1+R_2}\right)\dfrac{R_6}{R_5+R_6}}{\dfrac{R_6}{R_5+R_6}\dfrac{R_1}{R_2+R_1} - \dfrac{R_2}{R_2+R_1} - \dfrac{R_4}{R_4+R_3} - \dfrac{R_6}{R_5+R_6}} \tag{3.32}$$

As mentioned before, since passive elements have the task to adjust parameters like sensitivity and resolution and at the same time the working baseline frequency (readout speed), it is necessary to have a compromise between them. Due to the readout process, this interface is inherently robust with respect to low frequency noise, common mode disturbs, and parasitic capacitances.

3.3 Switchless Capacitance to Voltage Converters

The same approach based on a relaxation oscillator [13, 14] can be utilized to produce a voltage as output magnitude as in the solution reported in Figure 3.15.

$A1$ and $A2$ are the comparator and the integrator, respectively, that implement the relaxation oscillator, which is the core of the interface. In order to produce a voltage output, however, a differentiator ($A3$) and an inverting amplifier ($A4$) are added to the circuitry. The differential capacitance transducer is involved both in the integration (in particular, C_1) and in the differentiation (C_2) stages.

The working principle is the following. Voltages V_1 and V_2 are weighted by R_7 and R_8. Let us suppose that when the sum of the weighted V_1 and V_2

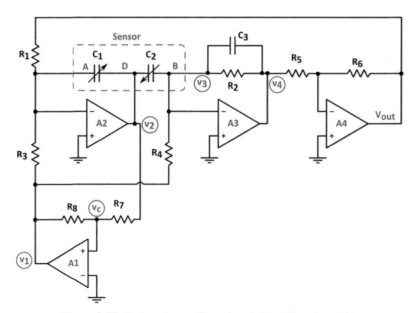

Figure 3.15   Relaxation oscillator-based C to V interface [13].

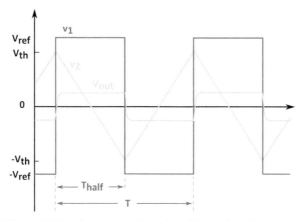

Figure 3.16   Time domain behavior of the analyzed interface.

(represented by V_c) is negative, the output of *A1* is negative; vice versa, when V_c is positive, the output of *A1* is positive (see Figure 3.16).

We can write that:

$$V_1 = V_{ref}\ sgn[V_c] \tag{3.33}$$

being:

$$V_c = V_1 \frac{R_7}{R_7 + R_8} + V_2 \frac{R_8}{R_7 + R_8} \tag{3.34}$$

and V_{ref} the saturation (supply) voltage.

Since the input of the integrator is a square wave, its output increases and decreases linearly with a triangular shape, while V_3 is a square wave.

Since C_3 is added for stability purposes, it can be neglected in the calculation of the interface output voltage. Considering all the Op-Amps as ideal, it is possible to write:

$$V_{out} = \frac{C_1 \frac{R_2}{R_4} - C_2 \frac{R_2}{R_3}}{C_1 \frac{R_5}{R_6} + C_2 \frac{R_2}{R_1}} V_{ref}\ sgn[V_c] \tag{3.35}$$

Equation (3.35) can be further simplified assuming that $R_3 = R_4$ and $R_1 R_5 = R_2 R_6$ as follows:

$$V_{out} = V_0\ sgn[V_c] \tag{3.36}$$

$$V_o = \frac{R_1}{R_3} \frac{C_1 - C_2}{C_1 + C_2} V_{ref} = kx V_{ref} \tag{3.37}$$

being k equal to R_1/R_3.

Therefore, the interface is capable of converting the *x* value into a linearly proportional voltage. Noticeably, in the case of linear sensors, the output period is independent from the measurand *x*, in fact:

$$T_o = 2(C_1 + C_2)R_3 \frac{R_7}{R_8} \tag{3.38}$$

Concerning the accuracy of the interface, there are two main sources of errors: passive element mismatches and nonideality of the Op-Amps. Both of them can be incorporated in the output expression as nonlinear and offset errors:

$$V_o = kx(1 + \varepsilon_R + \varepsilon_A + \varepsilon_V)V_{ref} + \Delta V_A + \Delta V_V + \Delta V_R \tag{3.39}$$

The terms ε_R, ε_A, and ε_V are the nonlinear errors coming from resistor mismatches, finite gain of the Op-Amps and their offset voltages, respectively, whereas $\Delta V_A, \Delta V_V$, and ΔV_R are the offset errors given by the same elements. By defining A the gain of the amplifiers, and V_{os} the offset voltage of the same devices, it is possible to define each of the sources of error in Equation (3.39) as follows:

$$\varepsilon_R = \frac{\left(\frac{R_3}{R_4} - 1\right) + \left(\left(\frac{R_1}{R_2} - 1\right) + \left(\frac{R_5}{R_6} - 1\right)\right)(1 + x)}{2}$$

$$\varepsilon_A = \frac{2}{A}(1 + x)$$

$$\varepsilon_V = \frac{R_3}{R_2}\frac{V_{os}}{V_{ref}} \tag{3.40}$$

$$\Delta V_A = \frac{V_{ref}}{A}$$

$$\Delta V_V = \left(1 + R_1\left(\frac{1}{R_2} + \frac{1}{R_3}\right)\right)V_{os}$$

$$\Delta V_R = \frac{\left(\frac{R_3}{R_4} - 1\right)kV_{ref}}{2}$$

A different approach is shown in the last interface presented in this section [15]. It is based on the synchronous demodulation technique and takes advantage of the possibility to work in feedback mode so to increase the output dynamic range (acting on the amplitude of the excitation signal). The overall processing takes place partially in the current domain and partially in

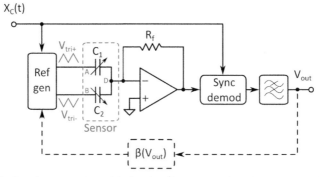

Figure 3.17 Synchronous demodulation-based C to V interface in open-loop conditions [14].

the voltage one, but we have inserted this solution in this section because the final demodulation and readout is performed in voltage mode.

The open-loop operation, whose schematic is shown in Figure 3.17, is based on a triangular wave generator, a transimpedance amplifier (TIA), a synchronous demodulator, and a low-pass filter. Capacitors C_1 and C_2 represent the equivalent sensor. The triangular wave is generated through a clock signal which drives four switches that, in turn, charge and discharge two reference capacitors C_t, as shown in Figure 3.18. The principle of operation is that the sensor is charged by the reference triangular voltages, while the TIA detects and amplifies the difference between the two currents flowing through the sensor. The switching synchronous demodulator is responsible for generating an output square wave where the information about the difference between the two capacitors is translated to the baseband. The demodulator is driven by the same clock signal which produces the reference triangular waves. A low-pass filter removes the unwanted high frequencies from the output spectrum. The output voltage as a function of the frequency is given by:

$$V_{out}(s) = 2\Delta C f_c R_f v_{in0} H_{LPF}(s) \tag{3.41}$$

where v_{in0} is the amplitude of the reference input signal, f_c is its frequency, and H_{LPF} is the transfer function of the low-pass filter at the output.

The sensitivity of this interface at low frequencies is given by:

$$S_{\Delta C}^{V_{out}} = \frac{dV_{out}}{d\Delta C} = 2R_f f_c v_{in0} \tag{3.42}$$

This interface is capable of working in closed loop mode as depicted in Figure 3.18. This allows to achieve a high dynamic range. The idea is to

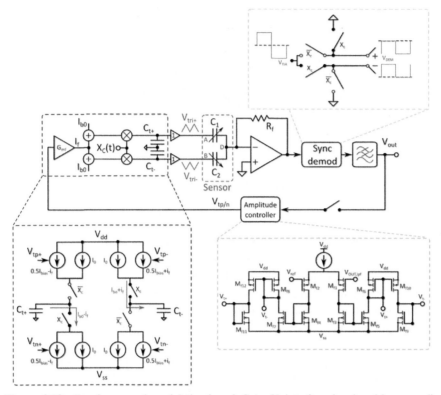

Figure 3.18 Synchronous demodulation-based C to V interface in closed-loop conditions [15].

evaluate the amplitude of the output signal to suitably tune the amplitude of the reference voltages through a controller module. In particular, this block adjusts the amplitude of the reference triangular signals (V_{tri+} and V_{tri-}) so that the larger capacitor is excited by a lower amplitude (this situation is depicted in Figure 3.19).

Once the system works in closed-loop mode, stability has to be ensured. By supposing a second-order low-pass filter response, the transfer function of the loop can be computed as:

$$H(s) = \frac{2Gm_f R_f C_0}{C_t} R_f H_{LPF}(s) \tag{3.43}$$

where Gm_f is the transconductance gain of the voltage to current converter depicted on the feedback path of Figure 3.16. A rule of thumb that grants a phase margin greater than 60° is that $C_0 \cong 3C_t$.

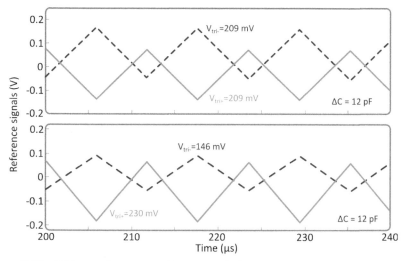

Figure 3.19 Difference between exciting waves during an open-loop operation (top) and a closed-loop (bottom) operation, supposing that C1 is greater than C2.

The sensitivity of the closed-loop interface can also be evaluated as:

$$S_{\Delta C}^{V_{out}} = \frac{dV_{out}}{d\Delta C} = \frac{2R_f I_{b0}}{C_t + 2C_0 R_f G_{mf}} \tag{3.44}$$

From the previous equation, it is clear that the closed-loop sensitivity depends on C_0 and C_t in agreement with the fact that the closed-loop mode adjusts itself according to the sensor values.

Lastly, the dynamic range of the interface can be evaluated as follows:

$$DR = \frac{\Delta C_{max}}{\Delta C_{min}} = \left(\frac{1}{2} + \frac{C_0}{C_t} R_f G_{mf}\right) \frac{V_{omax}}{v_{onoise}} \tag{3.45}$$

This value depends on the maximum output voltage V_{omax} (which is determined by the topology and the technology of the interface) and by the total output voltage noise (which limits the lower detectable value ΔC_{min}), expressed as an RMS value over the working bandwidth of the output LPF.

3.4 Capacitance to Digital Interfaces

In the last section of this chapter, two examples will be given on how it is possible to achieve a differential capacitance to digital conversion. The first

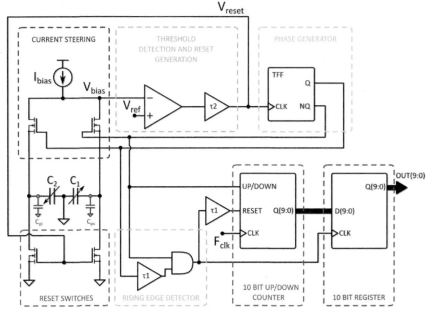

Figure 3.20   Capacitance to digital interface using a digital counter [16].

interface is shown in Figure 3.20 [16]. The differential sensor is embedded between two complementary rails (i.e., they cannot be simultaneously active) driven by a reference current I_{bias}. The readout process relies on the linear increment of the voltage V_{bias} caused by the constant current that charges either C_1 or C_2. The time that it takes for the voltage V_{bias} to reach the reference voltage of the comparator, V_{ref}, is determined by the up/down digital counter. Only one of the two rails is active and this is determined by the phase generator output driving Q_1 and Q_2 gates.

The idea is to count upwards during the charging of C_1 and downwards during the charging of C_2.

The difference between these two times allows to evaluate the difference between C_1 and C_2.

In particular, the readout process consists of two stages and starts when the signal Q coming from the phase generator is logically low (and therefore NQ is logically high). In this condition, the counter increases its digital output by one LSB at each clock cycle, the current I_{bias} starts charging C_1 through Q_2 (which is o,n) and V_{bias} raises up. This process goes on as long as V_{bias} is lower than V_{ref}. At that time, in fact, V_{reset} is pulled up, the sensor capacitors are discharged, the phase generator toggles its output, and the counter is set to count downwards. This transition does not influence the rising edge detector, and therefore, the counter output is not reset and starts counting backwards from the previous value. The total amount of time taken by this first phase is given by:

$$T_1 = \frac{C_1 V_{ref}}{I_{bias}} \tag{3.46}$$

Capacitor C_2 is charged during the second phase, which, again, lasts until the voltage V_{bias} (which is reverted to the *low* state during the reset pulse) goes higher than V_{ref}. When this condition is reached, the output of the phase generator toggles again restoring the starting condition. The total amount of time taken by the second stage is:

$$T_2 = \frac{C_2 V_{ref}}{I_{bias}} \tag{3.47}$$

At the output of the counter, however, there is a digital code representing the difference between T_1 and T_2 which is equal to T_{out}:

$$T_{out} = \frac{(C_1 - C_2) V_{ref}}{I_{bias}} \tag{3.48}$$

Since, at the end of the second stage, the signal Q of the phase generator toggles from high to low, the output of the rising edge detector spikes high, resetting the counter and storing the result of the readout process in the register. These two phases repeat indefinitely.

The conversion speed is given by the sum of T_1 and T_2:

$$T_{conv} = \frac{(C_1 + C_2) V_{ref}}{I_{bias}} \tag{3.49}$$

while the desired resolution imposes constraints on the number of bits N of the counter (Equation (3.50)), as well as on the clock frequency (Equation (3.51)).

$$V_{LSB} = \frac{V_{ref}}{2^N} \tag{3.50}$$

$$f_{clk} = \frac{2^N I_{bias}}{(C_1 + C_2) V_{ref}} \tag{3.51}$$

Reference current stability is fundamental for the accuracy of the sensor evaluation. Moreover, the presented design remains susceptible to the effects of parasitic capacitances.

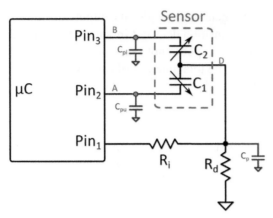

Figure 3.21   Direct microcontroller-based interface [17].

The second approach here described [17, 18] uses the microcontroller as a direct interface for the differential capacitive sensor, without adding any external circuitry. The configuration is shown in Figure 3.21. As visible, the sensor is connected to three digital terminals of the microcontroller and is represented together with its stray capacitances. Due to the reconfigurability of the pins where the sensor is connected, it is possible to change its boundary conditions, measure a number of different charge and discharge times through the microcontroller internal timers (according to the configuration), and, based on them, extract the value of the sensor itself.

The working principle is the following: by outputting a logical high (V_{dd}), pin 1 charges the total capacitance, C_t, which is a combination of the sensor and parasitic capacitors depending on the status of pin 2 and pin 3, through the resistor R_i. Once the total capacitance is fully charged, pin 1 changes its state to become an input terminal (high impedance) and measures the time that the equivalent capacitor C_t takes to discharge to a predetermined threshold (V_{TL}) through resistor R_d. To do that, the microcontroller uses its internal Schmitt trigger and counter (see Figure 3.22).

The procedure described so far is carried out three times, and during each time, pin 2 and pin 3 assume a different status. The first measurement is performed by outputting a logical 0 (gnd) on pin 2, while leaving pin 3 as a high impedance. The total capacitance given by this configuration is:

$$C_{t,1} = C_p + C_1 + \frac{C_2 C_{pl}}{C_2 + C_{pl}} \tag{3.52}$$

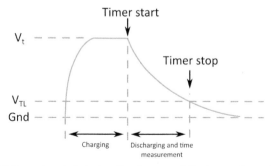

Figure 3.22 Voltage at node 1 during the charge and discharge phases driven by pin 1.

The discharge time is given by:

$$T_1 = R_d C_{t,1} \ln \left(\frac{V_{dd}}{V_{TL}} \right) \tag{3.53}$$

During the second measurement, the state of pin 2 and pin 3 is inverted; therefore, the former is set to high impedance, while the latter is set to zero. The total capacitance and discharge time can be calculated respectively as:

$$C_{t,2} = C_p + \frac{C_1 C_{pu}}{C_1 + C_{pu}} + C_2 \tag{3.54}$$

$$T_2 = R_d C_{t,2} \ln \left(\frac{V_{dd}}{V_{TL}} \right) \tag{3.55}$$

The last measurement is carried out with both pin 2 and pin 3 set to high impedance. Therefore, we can write that:

$$C_{t,3} = C_p + \frac{C_1 C_{pu}}{C_1 + C_{pu}} + \frac{C_2 C_{pl}}{C_2 + C_{pl}} \tag{3.56}$$

$$T_3 = R_d C_{t,3} \ln \left(\frac{V_{dd}}{V_{TL}} \right) \tag{3.57}$$

Once these three steps are concluded, the microcontroller performs the following operations to evaluate the sensor:

$$x = \frac{T_1 - T_2}{T_1 + T_2 - 2T_3} \tag{3.58}$$

Indeed, by substituting Equations (3.53), (3.55), and (3.57) into Equation (3.58), we can write:

$$x = \frac{C_1 \left(1 - \frac{C_{pu}}{C_1 + C_{pu}}\right) - C_2 \left(1 - \frac{C_{pl}}{C_2 + C_{pl}}\right)}{C_1 \left(1 - \frac{C_{pu}}{C_1 + C_{pu}}\right) + C_2 \left(1 - \frac{C_{pl}}{C_2 + C_{pl}}\right)} \qquad (3.59)$$

As visible, the Equation (3.59) is independent from V_{dd}, V_{TL}, and R_d, and therefore, the interface is insensitive to variations in these parameters. It is also noticeable that the capacitance C_p does not affect the readout, while C_{pl} and C_{pu} do not constitute an error source only if C_2 and C_1 are much greater than them.

References

[1] H. Barthelemy, "Current mode and voltage mode: Basic considerations," *46th Midwest Symposium on Circuits and Systems*, Cairo, 2003, pp. 161–163, Vol. 1.

[2] B. Gilbert, "Current mode, voltage mode, or free mode? A few sage suggestions," *Analog Integrated Circuits & Signal Processing*, 38(23), 83–101, 2004.

[3] K. Watanabe, S. Ogawa, Y. Oisugi and K. Kondo, "A switched-capacitor interface for differential capacitance transducers," *IMTC/99. Proceedings of the 16th IEEE Instr. and Meas. Technology Conference (Cat. No. 99CH36309)*, Venice, 1999, pp. 315–319, Vol. 1.

[4] S. Ogawa, Y. Oisugi, K. Mochizuki and K. Watanabe, "A switched-capacitor interface for differential capacitance transducers," *IEEE Transactions on Instrumentation and Measurement*, 50(5), 1296–1301, 2001.

[5] K. Nagaraj, T. Viswanathan, K. Singhal and J. Vlach, "Switched-capacitor circuits with reduced sensitivity to amplifier gain," *IEEE Transactions on Circuits and Systems*, 34(5), 571–574, 1987.

[6] B. George and V. J. Kumar, "Switched capacitor signal conditioning for differential capacitive sensors," *IEEE Transactions on Instrumentation and Measurement*, 56(3), 913–917, 2007.

[7] J. Shiah, H. Rashtian and S. Mirabbasi, "A low-noise parasitic-insensitive switched-capacitor CMOS interface circuit for MEMS

capacitive sensors," 2011 IEEE 9th International New Circuits and systems conference, Bordeaux, 2011, pp. 470–473.

[8] J. Shiah and S. Mirabbasi, "A 5-V 290-μW low-noise chopper-stabilized capacitive-sensor readout circuit in 0.8-μm CMOS using a correlated-level-shifting technique," *IEEE Transactions on Circuits and Systems II: Express Briefs*, 61(4), 254–258, 2014.

[9] X. Li, J. Hu and X. Liu, "A high-performance digital interface circuit for a high-Q micro-electromechanical system accelerometer." *Micromachines*, 9, 675, 2018.

[10] J. Shiah and S. Mirabbasi, "A 5-V 555-μW 0.8-μm CMOS MEMS capacitive sensor interface using correlated level shifting," 2013 IEEE International Symposium on Circuits and Systems (ISCAS), Beijing, 2013, pp. 1504–1507.

[11] K. Mochizuki, K. Watanabe, T. Masuda and M. Katsura, "A relaxation-oscillator-based interface for high-accuracy ratiometric signal processing of differential-capacitance transducers," IEEE Instrumentation and Measurement Technology Conference Sensing, Processing, Networking. IMTC Proceedings, Ottawa, Ontario, Canada, 1997, pp. 1164–1168, Vol. 2.

[12] A. De Marcellis, C. Reig and M.-D. Cubells-Beltrán, "A capacitance-to-time converter-based electronic interface for differential capacitive sensors," *Electronics*, 8, 80, 2019.

[13] K. Mochizuki, K. Watanabe and T. Masuda, "A high-accuracy high-speed signal processing circuit of differential-capacitance transducers," *IEEE Transactions on Instrumentation and Measurement*, 47(5), 1244–1247, 1998.

[14] G. Ferri et al. "Voltage-mode analog interfaces for differential capacitance position transducers". *Lecture Notes in Electronics Engineering*, 431, 2018.

[15] F. Aezinia and B. Bahreyni, "An interface circuit with wide dynamic range for differential capacitive sensing applications," *IEEE Transactions on Circuits and Systems. II: Express Briefs*, 60(11), 766–770, 2013.

[16] T. G. Constandinou, J. Georgiou and C. Toumazou, "A micropower front-end interface for differential-capacitive sensor systems," 2008 IEEE International Symposium on Circuits and Systems, Seattle, WA, 2008, pp. 2474–2477.

[17] F. Reverter and O. Casas, "Direct Interface Circuit for Differential Capacitive Sensors," IEEE Transactions on Instrumentation and Measurement Technology Conference, Victoria, 2008, pp. 1609–1612.

[18] F. Reverter and O. Casas, "Interfacing differential capacitive sensors to microcontrollers: A direct approach," *IEEE Transactions on Instrumentation and Measurement*, 59(10), 2763–2769, 2010.

4

Current Mode Differential Capacitive Sensors Interfaces

In analogy with the previous chapter, here we analyze how it is possible to interface a differential capacitance sensor with an analog circuit based on current mode approach. Unlike for the voltage mode, where the meaning of the definition was broader, there are not many works in the literature concerning current mode readout techniques, and almost all of them can be reconducted to a basic theory which will be introduced in the following pages.

Nevertheless, current mode operation has many benefits with respect to the voltage mode; specifically, it allows to achieve good dynamic range maintaining very low voltage levels. Moreover, current mode circuits are inherently simpler than voltage mode ones allowing a better integration and higher readout frequencies. On the other side, one of their main drawbacks is the sensitivity to parasitic capacitances that need, possibly, a compensation as shown in the interfaces analyzed in this chapter.

Then, we will investigate both "standard" devices solutions, typically integrated, since operations with currents at integrated level are relatively simple thanks to current mirror structures, and "nonstandard" approaches based on active devices like the second-generation current conveyors (CCIIs) and the second-generation voltage conveyors (VCIIs). In this sense, an appendix will be given at the end of the book to introduce the basic theory of these two blocks, useful for the reader to better understand also the behavior of the interfaces.

4.1 Current Mode Theory of Operation

Let us take into consideration Figure 4.1. It depicts the ideal configuration to approach a differential capacitive sensor in a current mode fashion. C_1 and C_2 represent the differential capacitive sensor, while I_{ref} is a constant reference

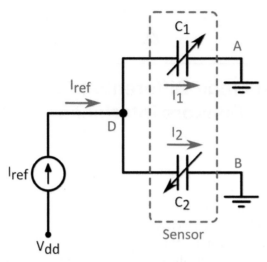

Figure 4.1   Ideal current mode approach.

current injected into the sensor itself. The injected current will be split into two portions I_1 and I_2 which depend on the actual value of C_1 and C_2. By performing the subtraction between the two currents, it is possible to evaluate the ratiometric parameter x as follows:

$$I_1 - I_2 = \frac{C_1}{C_1 + C_2} I_{ref} - \frac{C_2}{C_1 + C_2} I_{ref} = I_{ref} x \qquad (4.1)$$

Noticeably, a peculiarity of the current mode approach is that the sensitivity of the interface can be easily tuned by adjusting the reference current itself. This obviously has a drawback: an increment in the sensitivity of the interface comes at the expenses of an increased power consumption.

Figure 4.2 depicts a more realistic picture of this same approach, where parasitic capacitances have also been added. C_{pu} and C_{pl} give a negligible contribution for this analysis, given that they are grounded (the actual interface has to guarantee this condition in order for this analysis to be valid).

As a consequence, we can rewrite currents I_1 and I_2 as:

$$I_1 = \frac{C_1}{C_1 + C_2 + C_p} I_{ref}$$

$$\qquad (4.2)$$

$$I_2 = \frac{C_2}{C_1 + C_2 + C_p} I_{ref}$$

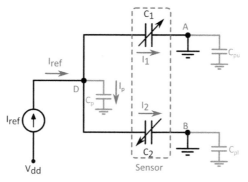

Figure 4.2 Current mode approach in the presence of parasitic capacitances.

Therefore, their difference results equal to:

$$I_1 - I_2 = \frac{C_1}{C_1 + C_2 + C_p} I_{ref} - \frac{C_2}{C_1 + C_2 + C_p} I_{ref}$$

$$= I_{ref} \frac{C_1 + C_2}{C_1 + C_2 + C_p} x \tag{4.3}$$

Recalling from Chapter 2 that the sum of the actual values of C_1 and C_2 for linear sensors remains constant regardless of the measurand:

$$C_1 + C_2 = 2C_{bl} \tag{4.4}$$

where C_{bl} represents the baseline (or standing value) of the sensor, and noticing that, although the function of the measurand, the sum of C_1 and C_2 for hyperbolic sensors can be expressed as:

$$C_1 + C_2 = \frac{2C_{bl}}{1 - x^2} \tag{4.5}$$

we can rewrite Equation (4.3) as follows:

$$I_1 - I_2 = \frac{I_{ref}}{1 + \frac{C_p}{2C_{bl}}} x \tag{4.6}$$

$$I_1 - I_2 = \frac{I_{ref}}{1 + (1 - x^2)\frac{C_p}{2C_{bl}}} x \tag{4.7}$$

where Equation (4.6) stands for linear sensors and Equation (4.7) stands for hyperbolic ones. The conclusions given in Chapter 2 are also valid for the current mode approach: the presence of a parasitic capacitance at the terminal D of the sensor determines a constant reduction in the reference current for linear sensors (loss of sensitivity) and a reduction that is a function of the measurand itself in the case of hyperbolic sensors (with consequent distortion in the input–output relationship).

These effects can be negligible for high baseline sensors (thousands of pF, where $C_p/2C_{bl}- > 0$) but greatly affect the correct readout for small baseline sensors (few pF or fF range).

4.2 Standard Device Interfaces

The first current mode interface proposed in this chapter is shown in Figure 4.3 [1]. Although sensitive to parasitic capacitances, it is designed to remain extremely low in complexity and symmetrical so as to ensure high readout speeds, low power consumption and rejection of common mode disturbs and switching error.

It is composed by two switches placed in parallel with the actual sensor and an analog block which performs the subtraction of the two currents I_1 and I_2 (detailed in Figure 4.4). The injection of a constant reference current I_{ref} causes an indefinite increment of the voltage at the D terminal of the sensor, inducing its saturation. Switches are therefore activated by an external clock signal so to provide a continuous discharge of the sensor between each

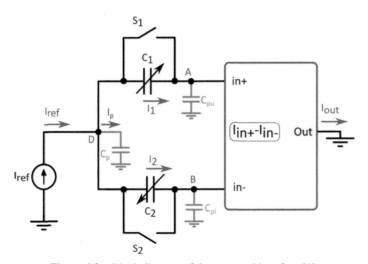

Figure 4.3   Block diagram of the proposed interface [1].

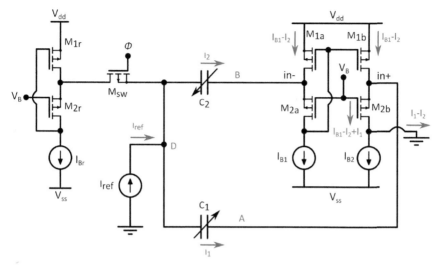

Figure 4.4 Actual implementation of the interface under analysis.

measurement. For this reason, the actual measurement consists of two phases: the discharge phase (T_D) and the readout phase (T_M). During the discharge phase, S_1 and S_2 are closed, so as to discharge the sensor and to give a zero current at the output, whereas once the switches are opened, the exact subtraction shown in Equation (4.3) is performed.

During the measuring phase, the voltage at the D node can be expressed as:

$$V_D = \frac{I_{ref}}{C_1 + C_2 + C_p} T_M \tag{4.8}$$

From this equation, it comes that it is possible to tune T_M and I_{ref} according to both the values of the sensor capacitors and the parasitic equivalent capacitance C_p.

The actual implementation of both the subtracting section and the switching one is proposed in Figure 4.4. Starting from the current subtracting section, the improved solution consists of a cascoded unitary current mirror $M_{1a,b}$, $M_{2a,b}$ whose DC voltage is set through the biasing voltage V_B. Fixed current sources I_{B1} and I_{B2} have ideally the same value. Given the negative feedback employed by M_{1a}, the virtual ground at the negative input is strong, whereas the same is not verified at the positive terminal. Therefore, the overall reference current has to remain below the bias current to preserve the virtual ground at that node.

The switching section consists of a single-pass transistor M_{sw} and a common gate section M_{2r} and M_{1r} offering low input impedance and mirroring M_{2a} and M_{1a}. In this way, when C_1 and C_2 are discharged, the charge stored in C_p is not transferred to the output, unlike how it is shown in Figure 4.3.

In this interface, both clock feedthrough and charge injection can be neglected: the first one, as already stated in Chapter 3, depends on the ratio between the overlap capacitance of the switching transistors and the total capacitance seen by the switching transistors themselves. The former obviously depends on technology parameters, but in general, being here a single switch solution, the overlap capacitance of M_{sw} is considerably lower than the sum of C_1, C_2, and C_p. Therefore, the clock amplitude transferred to the sensor tends to zero. Charge injection can be also neglected because, during switching phases, injected charges are drained by the low impedance M_{1r}, M_{2r} path. The overall main inaccuracy source is given by mismatches between paired transistors. This interface can be profitably used in linear sensors where the gain error is constant regardless of the measurand, or in high baseline hyperbolic sensors where the effects of stray capacitances can be neglected a priori.

Figure 4.5 shows another transistor level topology [2, 3] which, when applied to differential capacitive sensors, is capable of measuring their value by performing the difference between the two currents I_1 and I_2.

It consists of two pairs of complementary transistors driven by two clocks with the same frequency, same phase but different duty cycles so that only one transistor per branch is conducting at any given time. C_1 and C_2 are the sensor capacitors. In this topology, measurements are still affected by parasitic components, that is why, also in this case, linear sensors are still preferable to be used.

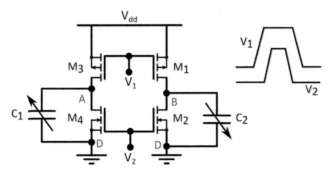

Figure 4.5   Simplified topology of the interface proposed in Ref. [2].

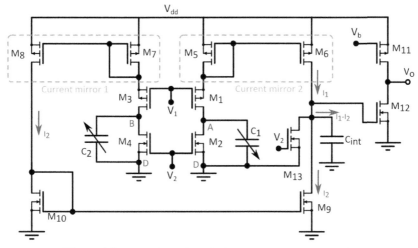

Figure 4.6 Full schematic with the I to V converter stage [2].

From a straightforward observation of the topology, it is possible to conclude that when V_1 and V_2 are high, M_2 and M_4 are conducting and C_1 and C_2 are grounded, hence discharged. On the contrary, when both V_1 and V_2 are low, M_1 and M_3 are "on," and therefore, the current flows through the sensor. The complete interface is shown in Figure 4.6. Other than the difference between the two currents, it also performs the current to voltage conversion. Current mirrors M_7, M_8 and M_5, M_6, similar to a symmetrical OTA, perform the sensing of I_1 and I_2 amplifying them according to the ratio $(W/L)_{M6,8}/(W/L)_{M5,7}$. The integrator capacitor C_{int}, together with the voltage buffer M_{12}, guarantees the current to voltage conversion. It can be evaluated as:

$$V_o = \frac{C_1 - C_2}{C_{int}} A_i V_{dd} + V_{DC} \tag{4.9}$$

where A_i is the actual gain of the current mirrors and V_{DC} is the DC voltage at the output of the interface.

It is worth to notice that although not directly considered into Equation (4.9), a parameter that affects the sensitivity of the interface is the actual geometry of transistors M_9 and M_{10} since a variation in their length changes consequently their output impedance. Also, since the interface relies on current mirrors, the accuracy limit of the interface can be reduced, other than by parallel parasitic capacitances, also by mismatches between paired transistors.

A more advanced interface capable of dynamically compensating parasitic capacitances is shown in Figure 4.7 [4]. It can be thought of as a direct evolution of the previous one. The sensor is again represented together with the parasitic capacitance at terminal D and is connected to a fully analog processing block that, thanks to its low input impedance, provides a virtual ground at its inputs, ideally nullifying stray capacitances at nodes A and B of the sensor. The analog block has two high impedance outputs where the sum and the difference of the input currents are produced.

The transconductance amplifier is demanded to generate a current which, once fed back to the sensor, provides a compensation for that portion of the reference current that is "stolen" from C_p to the actual sensor itself.

There are three switches (S_1, S_2, S_3) driven by two different clocks $(\Phi_1$ and Φ_2, see Figure 4.7). Both the clocks share the same frequency $(1/T)$ but different duty cycles. According to the switches status, the sensor readout is carried out in three steps, namely discharge (T_D), autotune (T_A), and measure (T_M); the overall period of the reference signal is therefore $T = T_D + T_A + T_M$.

During the discharge phase, switches are all closed. This ensures that the sensor is fully grounded (hence gets discharged) and that the transconductance amplifier output is equal to zero due to the fact that both its inputs are grounded as well. In other words, this is a reset state and makes sure that, at each cycle (T), the starting condition remains unchanged.

The autotune phase (evidenced in Figure 4.8) starts when switches S_1 and S_2 turn open. Current starts flowing through C_1, C_2, and C_p. Ideally $(C_p = 0)$, the sum of I_1 and I_2 is equal to I_{ref}, but, due to the presence of C_p, part of this current (I_p) is absorbed by the same capacitance. By looking then at the summing output and by choosing I_{comp} equal to I_{ref}, $I_1 + I_2$ is different from I_{comp} and the voltage across R_0 changes accordingly. This value is stored in C_h. The inputs of the transconductor (feedback controller) become of different values, and a current proportional to the "error signal" starts flowing from its output into C_p.

The measure phase starts when the switch S_3 opens. The resulting voltage coming from the autotune phase is held by C_h, while the total amount of current that drives the sensor is equal to:

$$I_{TOT} = I_{ref} + G_M R_O (I_{COMP} - I_1 - I_2) \tag{4.10}$$

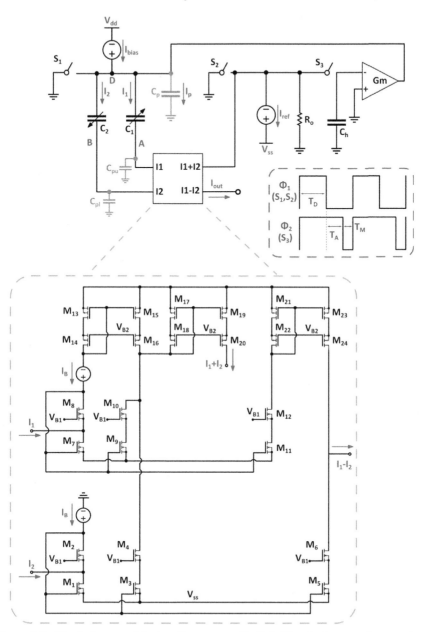

Figure 4.7 Block diagram of the interface proposed in Ref. [4] and a focus on the summing-subtracting block and clock signal timings.

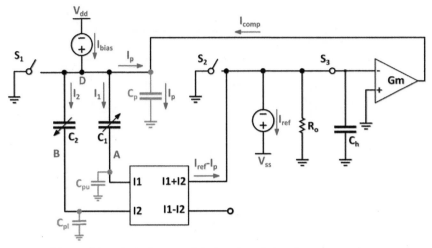

Figure 4.8   Block diagram of the interface during the autotune phase.

Currents I_1 and I_2 can be expressed as follows:

$$I_1 = (I_{ref}(1 + A) - A(I_1 + I_2))\frac{C_1}{C_T} \tag{4.11}$$

$$I_2 = (I_{ref}(1 + A) - A(I_1 + I_2))\frac{C_2}{C_T} \tag{4.12}$$

where, in the previous equations, I_{COMP} is replaced by I_{ref}, $C_1 + C_2 + C_p$ is equal to C_T and A is equal to $G_M R_O$.

Since the sum of I_1 and I_2 can be expressed as:

$$I_1 + I_2 = I_{ref}\frac{(1 + A)(C_1 + C_2)}{C_T + A(C_1 + C_2)} \tag{4.13}$$

their difference can be consequently calculated (using Equations (4.4) and (4.11–4.13)) as:

$$I_1 - I_2 = \frac{I_{ref}2C_{bl}}{\frac{C_T}{1+A} + 2C_{bl}}x \tag{4.14}$$

for linear sensors, while (using Equations (4.5) and (4.11–4.13)) as:

$$I_1 - I_2 = \frac{I_{ref}}{1 + (1 - x^2)\frac{C_p}{2C_{bl}(1+A)}}x \tag{4.15}$$

for hyperbolic ones. In both the cases, when A tends to be infinite, the equation tends to be the ideal one (Equation (4.1)).

The frequency of the clock signal as well as the amplitude of the reference current can be set according to the maximum voltage that the designer wants the node D to reach, in particular:

$$V_D = \frac{I_{ref} + I_{GM}}{C_1 + C_2 + C_p} T_M \qquad (4.16)$$

Obviously, for a higher reference current, the measurement can be faster. Increasing the reference current improves the resolution of the interface; even if power consumption increases, the comparison current I_{COMP} must be considered.

4.3 CCII- and VCII-based Interfaces

In this paragraph, two examples on how it is possible to approach differential capacitance sensors interfacing using either second-generation current conveyors (CCIIs) or second-generation voltage conveyors (VCIIs) will be shown. Although constitutive relationships are given in the following lines, Appendix will give a deeper overview about these two active current mode blocks.

A CCII-based interface is shown in Figure 4.9 [5]. The input–output relationship of such a device is given by the following matrix [6–10]:

$$\begin{bmatrix} I_Y \\ V_X \\ I_Z \end{bmatrix} = \begin{bmatrix} 0 & 0 & 0 \\ \alpha & 0 & 0 \\ 0 & \pm\beta & 0 \end{bmatrix} \begin{bmatrix} V_Y \\ I_X \\ V_Z \end{bmatrix} \qquad (4.17)$$

Parameters α and β should be close to unity, so that the voltage on the Y terminal is conveyed to the X terminal, while the current on the X terminal is conveyed to the Z terminal. According to the verse of the currents on X and Z, it is possible to identify positive and negative current conveyors, that is why β has a sign.

The interface is composed by a CCII+ and a CCII−, while the differential capacitive sensor is connected to the low impedance X terminal of the active devices. An oscillating current source drives the common terminal D of the sensor. Given the conveying action between Y and X terminals, connecting the former to ground theoretically nullifies parasitic capacitances C_{pu} and C_{pl}, while C_p remains uncompensated. The difference at node O is automatically carried out thanks to the inversion of I_2 current which is given by the CCII− (refer to Figure 4.9). Capacitor C_3 adds an overall offset to the output

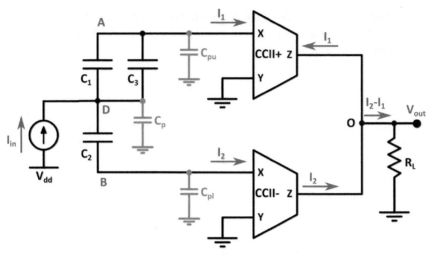

Figure 4.9 CCII-based differential capacitive sensor interface [5].

signal so that, if suitably dimensioned, it allows to translate the input–output relationship into positive voltages even for negative values of x.

The output is given in the form of voltage across the resistor R_L and, supposing ideal CCIIs, can be expressed as:

$$V_{out} = R_L I_{in,pp} \left(\frac{C_1 + C_2}{C_1 + C_2 + C_3} x + \frac{C_3}{C_1 + C_2 + C_3} \right) \quad (4.18)$$

The properties of the CCII can be profitably exploited to use an oscillating voltage rather than current in order to produce the reference signal, as shown in Figure 4.10. As visible, a third CCII is added. At its Y terminal, the reference voltage is connected; this is converted into a current at X terminal, thanks to the resistor R_1, and finally conveyed to the Z terminal for the readout.

The input–output relationship can be written as:

$$V_{out} = \frac{R_L V_{in,pp}}{R_1} \left(\frac{C_1 + C_2}{C_1 + C_2 + C_3} x + \frac{C_3}{C_1 + C_2 + C_3} \right) \quad (4.19)$$

Noticeably, since a negative CCII can be implemented by a pair of positive CCIIs (according to the configuration reported in Figure 4.11), at the expenses of an increment in area occupation, the interface can be implemented using a single active block.

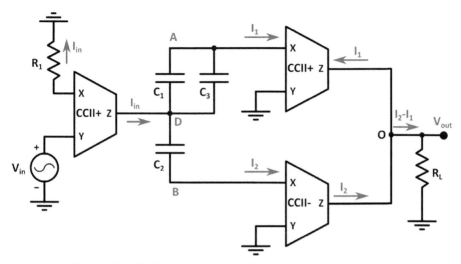

Figure 4.10 CCII-based interface with a reference voltage generator.

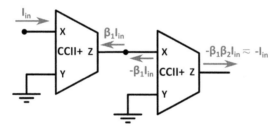

Figure 4.11 A negative CCII implemented by means of two positive CCIIs; it is supposed that the β parameter is close to unity so that it can be neglected in the input–output relationship.

A VCII-based interface is shown in Figure 4.12 [11]. As the dual device of the CCII, a second-generation voltage conveyor is a three-terminal device, whose input–output relationships can be extracted from the following matrix [12–16]:

$$\begin{bmatrix} i_x \\ v_y \\ v_z \end{bmatrix} = \begin{bmatrix} 0 & \pm\beta & 0 \\ 0 & 0 & 0 \\ \alpha & 0 & 0 \end{bmatrix} \begin{bmatrix} v_x \\ i_y \\ i_z \end{bmatrix} \tag{4.20}$$

The α and β parameters (also in this case ideally unity) represent, as for the CCII block, the ratio between voltages and currents, respectively. However, for a VCII, the current mirroring action is carried out between the Y and X terminals, $I_X = \pm\beta I_Y$, while the voltage at the X terminal is conveyed

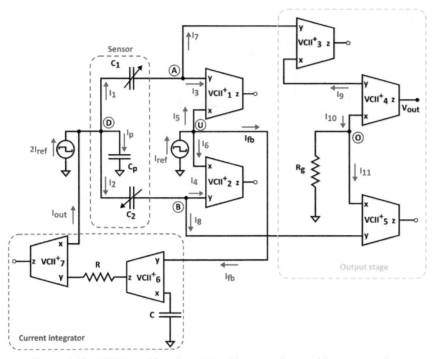

Figure 4.12   VCII-based interface [11] with automatic parasitic compensation.

to the output Z terminal $V_Z = \alpha V_X$. The Y terminal, due to its very low input impedance, can be considered at virtual ground ($V_Y = 0$).

From a designer perspective, a voltage conveyor results more versatile than a current conveyor.

The presented interface, other than the readout of the sensor, aims to dynamically compensate node D parasitic capacitance (C_p) using a time continuous current feedback approach.

The circuit can be divided into three parts: the *first part* is a VCII-based current summing-subtracting block composed of VCII$_1$–VCII$_2$ and a current source equal to I_{ref}. The sensor input is excited by a current source equal to $2I_{ref}$. Reference currents are square wave signal. By this solution, the charging and discharging of capacitors are automatically performed without any need to use switches. Virtual ground and low impedance at Y terminal of VCII keep the sensor second terminal to ground allowing the mitigation of C_{pu} and C_{pl} stray capacitances.

In this section, the sum of I_1 and I_2 is produced and then subtracted from I_{ref} to obtain I_{fb} (error signal), which, suitably integrated, generates the current taken by C_p (I_{out} tends to $-I_p$).

The *second part* is the control section, which feeds I_{fb} to the sensor input. It is a VCII-based current-input current-output integrator composed of the two VCII blocks, a resistor, and a grounded capacitor.

In the *third part* (output stage), the difference between I_1 and I_2 is produced and converted to a proportional voltage signal. This operation is performed by three VCII blocks (VCII$_3$–VCII$_5$) and resistor R_g.

Currents I_1 and I_2 can be written, respectively, as:

$$I_1 = \lambda(2I_{ref} - I_p) \tag{4.21}$$

$$I_2 = (1 - \lambda)(2I_{ref} - I_p) \tag{4.22}$$

Therefore, currents I_3, I_4, I_7, and I_8 can be expressed as:

$$I_3 = I_7 = \frac{I_1}{2} = \frac{\lambda}{2}(2I_{ref} - I_p) \tag{4.23}$$

$$I_4 = I_8 = \frac{I_2}{2} = \frac{1 - \lambda}{2}(2I_{ref} - I_p) \tag{4.24}$$

At node U, it is possible to write that:

$$I_{ref} = I_5 + I_6 + I_{fb} \tag{4.25}$$

Since $I_5 = \beta_1 I_3$ and $I_6 = \beta_2 I_4$, using Equations (4.23–4.25), we have:

$$I_{ref} = \beta_1 \frac{\lambda}{2}(2I_{ref} - I_p) + \beta_2 \frac{(1 - \lambda)}{2}(2I_{ref} - I_p) + I_{fb} \tag{4.26}$$

If β_1 and β_2 are equal to unity, Equation (4.26) shows that I_{fb} must be equal to $I_p/2$ for the feedback to reach the steady state.

In fact, at node U, a current comparing action is performed. As long as the sum of I_5 and I_6 is not equal to I_{ref} (i.e., $I_1 + I_2$ is not equal to $2I_{ref}$), a non-zero compensation current (I_{fb}) is produced. This current is fed to the input node by the controller circuit, which is a VCII-based current integrator. By assuming virtual ground at Y port of VCII, the relationship between the input and output currents of the integrator can be expressed as:

$$I_{out} = \alpha_7 \beta_7 \beta_8 \frac{1}{sRC} I_{fb} \tag{4.27}$$

Once the feedback circuit reaches the steady state (hence compensating the effects of C_p), the output voltage can be simply calculated as:

$$V_{out} = \alpha_4 V_B = \alpha_4 R_g (I_{10} - I_{11}) \tag{4.28}$$

But since I_{10} and I_{11} are, respectively, equivalent to:

$$I_{10} = \beta_4 I_9 = \beta_3 \beta_4 I_7 = \beta_3 \beta_4 \frac{I_1}{2} \tag{4.29}$$

$$I_{11} = \beta_5 I_8 = \beta_5 \frac{I_2}{2} \tag{4.30}$$

Assuming the involved α and β parameters equal to unity, it is possible to conclude that:

$$V_{out} \approx \frac{R_g(I_1 - I_2)}{2} = R_g I_{ref} x \tag{4.31}$$

Equation (4.31) shows a linear relationship between x and V_{out}. As a consequence, the sensitivity of the circuit is constant and regulated by adjusting I_{ref}, R_g, or both of them:

$$S_{\Delta C}^{Vout} = \frac{dV_{out}}{d\Delta C} = \frac{d\left(R_g I_{ref} \frac{\Delta C}{C_{bl}}\right)}{d\Delta C} = \frac{R_g I_{ref}}{C_{bl}} \left[\frac{V}{F}\right] \tag{4.32}$$

Finally, we underline that the design of the integrator (and, in general, of the feedback controller) determines the readout speed of the interface. In particular, its time constant RC must be sufficiently smaller than the half-period of the reference current I_{ref}, ensuring that the loop reaches the steady state before the reference signal changes polarity, consequently, completing an effective measurement.

References

[1] S. Pennisi, "High-performance and simple CMOS interface circuit for differential capacitive sensors," *IEEE Transactions on Circuits and Systems II: Express Briefs*, 52(6), 327–330, 2005.
[2] E. G. Zadeh and M. Sawan, "High accuracy differential capacitive circuit for bioparticles sensing applications," 48th Midwest Symposium on Circuits and Systems, 2005, Covington, KY, 2005, pp. 1362–1365, Vol. 2.

[3] D. Sylvester, J. C. Chen and C. Hu, "Investigation of interconnect capacitance characterization using charge-based capacitance measurement (CBCM) technique and three-dimensional simulation," *IEEE Journal of Solid-State Circuits*, 33(3), 449–453, 1998.

[4] G. Scotti, S. Pennisi, P. Monsurrò and A. Trifiletti, "88-uA 1-MHz stray-insensitive CMOS current-mode interface IC for differential capacitive sensors," *IEEE Transactions on Circuits and Systems I: Regular Papers*, 61(7), 1905–1916, 2014.

[5] G. Ferri, F. R. Parente and V. Stornelli, "Analog current-mode interfaces for differential capacitance sensing," 2016 IEEE Sensors Applications Symposium (SAS), Catania, 2016, pp. 1–6.

[6] K. Smith and A. Sedra, "The current conveyor—A new circuit building block," *Proceedings of the IEEE*, 56(8), 1368–1369, 1968.

[7] G. Ferri and N. Guerrini, *Low-voltage low-power CMOS current conveyors*. Boston, MA: Springer US, 2004.

[8] L. Grigorescu, "Amplifier built with current conveyors," *Romanian Journal of Physics*, 53, 109–113, 2008.

[9] V. Stornelli, G. Ferri, L. Pantoli, G. Barile and S. Pennisi, "A rail-to-rail constant-g_m CCII for Instrumentation Amplifier applications," *AEU – International Journal of Electronics and Communications*, 91, 103–109, 2018.

[10] J. A. Svoboda, "Current conveyors, operational amplifiers and nullors," *IEE Proceedings G – Circuits, Devices and Systems*, 136(6), 317–322, 1989.

[11] G. Barile, L. Safari, G. Ferri and V. Stornelli, "A VCII-based stray insensitive analog interface for differential capacitance sensors," *Sensors*, 19(16), 3545, 2019.

[12] J. Cajka and K. Vrba, "The voltage conveyor may have in fact found its way into circuit theory," *International Journal of Electronics and Communication (AEUE)*, 58, 244–248, 2004.

[13] I. M. Filanovsky, Current conveyor, voltage conveyor, gyrator, Proceedings of the 44th IEEE 2001 Midwest Symposium on Circuits and Systems, 1, 314–317, 2001.

[14] L. Safari, G. Barile, V. Stornelli and G. Ferri, "An overview on the second generation voltage conveyor: Features, design and applications," *IEEE Transactions on Circuits and Systems II: Express Briefs*, 66(4), 547–551, 2019.

[15] G. Barile, G. Ferri, L. Safari and V. Stornelli, "A new high drive class-AB FVF based second generation voltage conveyor," *IEEE Transactions on Circuits and Systems II: Express Briefs*, 2019.

[16] L. Pantoli, G. Barile, A. Leoni, M. Muttillo and V. Stornelli, "A novel electronic interface for micromachined Si-based photomultipliers," *Micromachines*, 9(10), 507, 2018.

5

Autobalanced Bridge-based Differential Capacitive Sensor Interfaces

In this chapter, we present how the autobalanced bridge technique can be effectively applied to the readout of a differential capacitive sensor. This technique has been also employed with both resistive bridges (i.e., Wheatstone topology) and, more generally, impedance bridges (for instance, De Sauty topology).

The idea behind this approach is the performance improvement of the basic bridge structure. In fact, it becomes suitable for unknown baseline sensors evaluation (uncalibrated interface) and the variation of the bridge elements greatly improves.

After a short introduction about the basic working principles of the autobalanced bridge technique, in the first part we will show a nonlinear readout circuit, as well as its linearization. In the second part, it is going to be introduced as a methodology based on a feedback loop that, working as a parallel structure with respect to the actual interface, can be employed in this scenario to mitigate the effects of parasitic capacitances.

5.1 The Autobalanced Bridge Architecture

Figure 5.1 shows a simple impedances bridge, excited by a sinusoidal reference voltage having an amplitude equal to V_{ref}. The equations that establish the balance condition are:

$$\begin{cases} |Z_1||Z_3| = |Z_2||Z_s| \\ <Z_1> + <Z_3> = <Z_2> + <Z_s> \end{cases} \quad (5.1)$$

where the upper and the lower identities indicate the balance condition for the magnitude and the phase (indicated as $<\cdot>$) of the impedances, respectively.

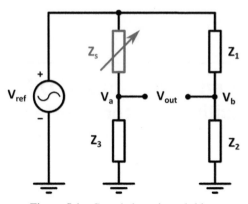

Figure 5.1 Generic impedance bridge.

Given that Equation (5.1) is verified, and the bridge stays at the balance condition, the output voltage is equal to zero:

$$V_{out} = V_a - V_b = 0 \tag{5.2}$$

From a designer perspective, however, a sensible solution to achieve this condition, disregarding for a moment the phase constraints, is to set all the elements equal in magnitude. Indeed, by doing this, it is ensured that the amplitude of the signals V_a and V_b at each bridge branches, at the steady state, is at half of the reference voltage amplitude V_{ref}. This ensures in fact the maximum readout dynamic range for the interface (see Figure 5.2).

Two issues are evidenced: first, to implement the condition coming from the previous lines, it is fundamental the knowledge of the sensor baseline, from which it is possible to choose the other bridge components. This means that in order to work adequately, the bridge has to undergo to a calibration phase. Note that sometimes shifting V_a and V_b to a value different from $V_{ref}/2$ is something wanted, for instance, if the sensor magnitude raises or lowers only. Nonetheless, it is required the knowledge of the sensor baseline.

The second issue is that even if properly calibrated, a bridge-based interface remains unsuitable for a largely variable measurand due to the fact that, being V_a or V_b (according to the position of the sensor) free to vary, if the sensor varies too much, in addition to output nonlinearity, one or the other voltages inevitably saturates. Moreover, the actual differential amplifier that produces the difference between the voltages V_a and V_b must have tight constraints as, for instance, high input and output dynamic range, in order to perform an accurate readout at any conditions.

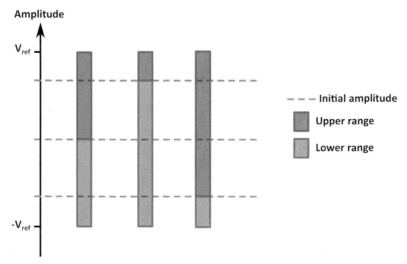

Figure 5.2 Dynamic ranges available at different values of impedances: All equal values (left); $Z_3 < Z_s$ and $Z_2 < Z_1$ (center); $Z_3 > Z_s$ and $Z_2 > Z_1$ (right).

Adding the autobalance feature to the bridge [1–6] allows to mitigate all the aforementioned issues: the basic method to implement such a structure is to add some circuitry to the bridge so as to implement a negative feedback, whose aim is to maintain V_a equal to V_b (and therefore V_{out} equal to zero) at any given measurand value.

Taking Figure 5.3 as a reference, the bridge is balanced through a voltage controlled impedance (VCI) that varies according to the sensor to maintain V_a fixed and equal to V_b. By doing so, as long as the voltage controlled impedance is capable of assuming the value that balances the bridge, it is possible to obtain the following advantages.

The first and very important one is that there is no need to know the sensor baseline; indeed, the bridge, due to the negative feedback, self-calibrates to the reference voltage given by the fixed branch (V_b in Figure 5.3) according to the values of Z_1 and Z_2, making the user able to perform the readout as long as the relationship between the control voltage and the voltage-controlled impedance value is known. The second advantage is that the voltage V_a is fixed to the value of V_b (which in turn is fixed at the value desired by the designer), avoiding its saturation to one extreme of the supply voltage. In other words, the span of the sensor that the interface is capable of reading is determined by the variation of the voltage-controlled impedance itself and not anymore by the voltage saturation.

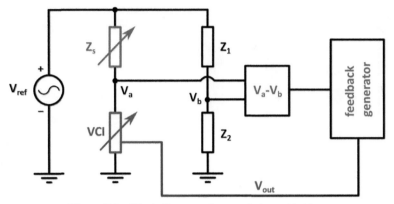

Figure 5.3 Block diagram of an autobalanced bridge.

The actual topology of the bridge as well as how to implement a voltage-controlled impedance depends on the designer and is going to be analyzed in the following pages.

The remaining part of the feedback produces the voltage to drive the voltage-controlled impedance, voltage that is also the actual output of the interface. Since that a synchronous demodulation process can be used to extract it from the difference (error signal) $V_a - V_b$, an autobalanced bridge interface also enjoys the inherent advantages coming from that procedure (noise reduction and offset error mitigation).

5.2 Nonlinear Output Interface

The block diagram of the first interface is shown in Figure 5.4 [7, 8]. As visible, the impedance bridge is arranged so that its left branch is formed by the differential capacitive sensor itself (C_1, C_2), while the right one by a fixed resistance (R) and a voltage-controlled resistance (VCR, named R_{vcr}) which is responsible for maintaining the bridge at its equilibrium by changing its value according to the differential capacitive sensor variation. This particular placement of the components allows to disregard the phase contribution in case of bridge unbalancing since both V_a and V_b are always in phase with V_{ref}.

The feedback path is composed by a differential amplifier, which allows to extract the difference between V_a and V_b (ΔV), and by an analog multiplier which, together with the integrator, have the dual task of both performing

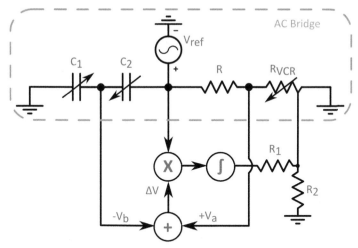

Figure 5.4 Autobalanced bridge interface proposed in Ref. [7].

the synchronous demodulation and generating the actual output voltage V_{ctrl}, that, in turn, is also the driving signal for the VCR.

Differential capacitive variations change V_b value in the left branch of the bridge. As a consequence, through the feedback loop, V_{ctrl} level is varied and R_{vcr} changes its value so as to force the bridge to be at the equilibrium. Noticeably, the signal ΔV, in a configuration like this, represents the "error signal" of the feedback architecture: the overall system indeed makes sure that it remains equal to zero.

To fully understand the working principle of the interface, its electronic implementation (shown in Figure 5.5) can be analyzed. A possible way to implement a voltage-controlled resistor is given by the use of an analog multiplier in the so-called Zhong configuration (see Figure 5.6). Supposing to use an AD633 as multiplier device, the equivalent resistance versus voltage relationship is given by:

$$R_{VCR} = \frac{10R_f}{10[V] - V_{ctrl}} \tag{5.3}$$

where [V] means that the unit of measurement is in Volts.

The voltage divider $R_1 - R_2$ is added to ensure that even if the VCR saturates (the bridge cannot reach the balance condition), the control voltage does not exceed the maximum value allowed by the VCR.

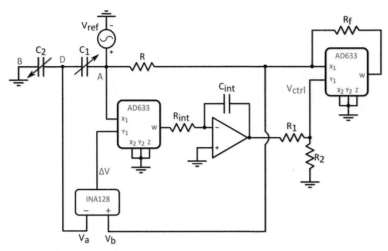

Figure 5.5   Circuit diagram of the analyzed interface.

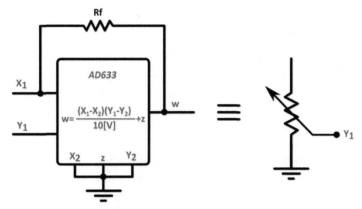

Figure 5.6   VCR implemented by means of a Zhong configuration.

Voltages V_a and V_b can be evaluated as:

$$V_a = V_{ref} \frac{R_{VCR}}{R + R_{VCR}} \tag{5.4}$$

$$V_b = V_{ref} \frac{\frac{1}{sC_2}}{\frac{1}{sC_1} + \frac{1}{sC_2}} = V_{ref} \frac{C_1}{C_1 + C_2} \tag{5.5}$$

By supposing to substitute to C_1 and C_2 their linear parametrization (Equation (2.13)), it is possible to evaluate the difference ΔV between

V_a and V_b as:

$$\Delta V = V_{ref} \left(\frac{R_{VCR}}{R + R_{VCR}} - \frac{1 + x}{2} \right) \tag{5.6}$$

Substituting Equation (5.3) into (5.6), we have:

$$\Delta V = V_{ref} \left(\frac{10}{10[V] - V_{ctrl}} - \frac{1 + x}{2} \right) \tag{5.7}$$

It is finally possible to manipulate Equation (5.7) obtaining the following expression of the x parameter as a function of V_{ref}:

$$x = \frac{V_{ctrl}}{20[V] - V_{ctrl}} - 2\frac{\Delta V}{V_{ref}} \tag{5.8}$$

The equation above is valid for the full range of x variation ($\pm 100\%$ even if the VCR saturates); indeed, it expresses the value of the measurand as a function of ΔV as well. If the VCR is able to guarantee the balance of the bridge, the equation can be reduced to the first term (see Figure 5.7) in the so-called autobalancing region:

$$x = \frac{V_{ctrl}}{20[V] - V_{ctrl}} \tag{5.9}$$

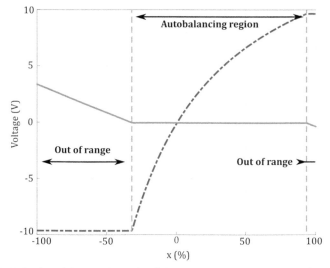

Figure 5.7 Behavior of the interface when it is in range (VCR compensates imbalances) and out of range (VCR saturated).

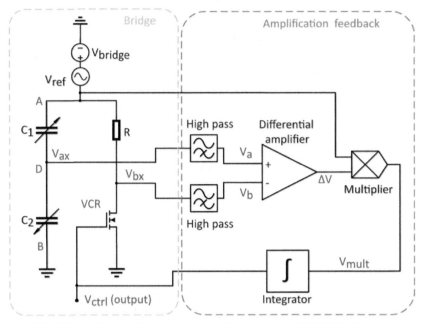

Figure 5.8   Differential capacitance sensor interface that uses a MOSFET in linear region as a VCR [9].

From Equation (5.9), it is evident how x can be easily determined only by reading V_{ctrl}. The overall readout speed depends on the settling time of the loop, which, in turn, is determined by the demodulation section. Since it is also the feedback controller, it has to be designed to fulfill the residual ripple constraints on the output signal.

The integrated version of this circuit can be easily implemented by using a MOSFET working in linear region as a VCR [9] as shown in Figure 5.8. Differently from the previous solution, it is explicitly put in evidence the DC component of the bridge. Indeed, bridge biasing (V_{bridge}) and resistor R have to be designed to allow the MOS VCR to work in linear region, ideally throughout the entire load line of the transistor, so as to maximize the auto-balance range of the interface. It is also possible to suitably choose the transistor dimensions, particularly increasing its length so as to guarantee a wider linear region. Differently from the solution in Ref. [7], high-pass filters are also added at the bridge output (refer to Figure 5.8) so as to avoid the DC component of the bridge to influence the following stage-biasing voltage.

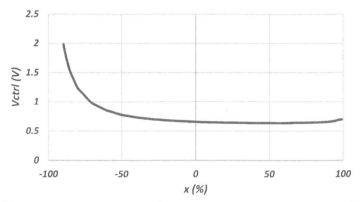

Figure 5.9 Input–output relationship of the analyzed NMOS-based VCR interface using generic technology parameters.

Starting back from Equation (5.6), and considering that the drain-source resistance of a MOSFET working in linear region is equal to:

$$R_{ds} = \frac{1}{\mu_n C_{ox} \left(\frac{W}{L}\right) (V_{gs} - V_{th} - V_{ds})} \tag{5.10}$$

it is possible to evaluate x as follows:

$$x = \frac{1 - R\mu_n C_{ox} \left(\frac{W}{L}\right) (V_{gs} - V_{th} - V_{ds})}{1 + R\mu_n C_{ox} \left(\frac{W}{L}\right) (V_{gs} - V_{th} - V_{ds})} \tag{5.11}$$

Both in Equations (5.10) and (5.11), μ_n represents the mobility of electrons in a silicon medium (supposing to use an NMOS), C_{ox} is the capacitance per unit of area for the technology used, W/L is the form factor of the transistor and V_{th} is the threshold voltage of the MOSFETS for the specific technology. For sake of clarity, an indicative plot of Equation (5.11) is given in Figure 5.9. As visible, both the interface versions respond to a measurand variation in a highly nonlinear fashion, so the interface depends on the actual measurand amplitude. Moreover, in both the cases, the VCR is not capable of maintaining the balance condition through the entire input full-range.

5.3 Linear Output Interface

In this section, an architecture that overcomes the aforementioned limitations of linearity and autobalancing region limitation is presented. The basic idea is here described: as stated before, the autobalancing strategy can be considered as a negative feedback-based system, whose aim is to minimize or null a certain error signal with the advantage that the input–output relationship only depends on the feedback gain.

According to well-known results of the automatic control theory, such a nulling effect can be obtained using a very high open-loop gain and integral control strategy. This technique can be used for bandpass signals as well (i.e., when signals of interest are phasors), as shown in Figures 5.5 and 5.8, but in order to use the integral control strategy, a baseband signal has to be generated. The solution here described [10, 11] uses therefore a synchronous demodulator in the forward path but unlike Figure 5.5, and Figure 5.8 (schematized in Figure 5.10) introduces a modulator in the feedback path, as depicted in Figure 5.11, allowing to move the signal from bandpass to baseband domains and vice versa. The intrinsic advantage is that a regular integral control strategy can still be applied on the baseband signal, whereas the VCR action only modifies the amplitude (see Figure 5.12) of the signal V_b, therefore generating a linear relationship between the control voltage V_{ctrl} and the x parameter. When an integral control is applied and both the forward

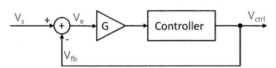

Figure 5.10   Schematic representation of the generic autobalancing system.

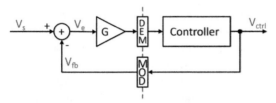

Figure 5.11   Schematic representation of the baseband versus bandpass signals in the analyzed interface [10].

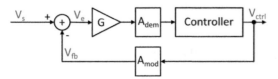

Figure 5.12   Simplification of the analyzed interface feedback circuit in the time domain [10].

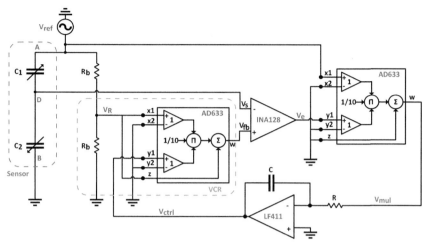

Figure 5.13 Schematic diagram of the analyzed interface [10].

(i.e., $A = G \cdot A_{mod} \cdot s^{-1}$ in Laplace domain) and feedback paths ($B = A_{dem}$) are linear, the closed loop transfer function is given by Equation (5.12). In other words, the system behaves as a first-order low-pass filter whose cutoff angular frequency is $\omega_{C0} = A \cdot B$.

$$H(s) = \frac{A \cdot \frac{1}{s}}{1 + A \cdot \frac{1}{s} \cdot B} = \frac{1}{B} \frac{1}{1 + s\frac{1}{A \cdot B}} \tag{5.12}$$

The whole interface circuit is shown in Figure 5.13.

In order to calculate its input–output relationship, the assumption of using AD633 as analog multiplier has been made.

The sensor, represented by capacitors C_1 and C_2, constitutes the left branch of the bridge excited by the sinusoidal AC source V_{ref}. At the rest position ($x = 0$), the sensor output is:

$$V_s(t) = \frac{V_{ref}(t)}{2} = V_R(t) \tag{5.13}$$

The VCR in the right branch is controlled in order to generate the V_{fb} signal, following the V_S. The rightmost dashed box of Figure 5.13 indicates the VCR actual implementation, based on fixed resistor R_b and an analog multiplier AD633.

The feedback signal summing node is implemented by an Instrumentation Amplifier, INA, providing buffering as well, being configured for a unitary gain. The synchronous demodulation of the error signal V_e is performed using another AD633. The multiplier output V_{mul} can be computed as:

$$V_{mul}(t) = \frac{1}{10\,[V]} \frac{|V_{ref}|\,|V_e|}{2}(1 - cos(2\omega_{ref}t)) \qquad (5.14)$$

Signal V_{mul} feeds an inverting integrator which, in turn, generates the signal V_{ctrl} as calculated in Equation (5.15). The final approximation holds when the residual ripple, due to the nonideal behavior of the integrator in removing unwanted high-frequency components of the demodulation process, is negligible. However, it can be removed by further post-processing. The control voltage V_{ctrl} is given by:

$$\begin{aligned}
V_{ctrl}(t) &= -\frac{1}{RC}\frac{|V_{ref}|}{20[V]}\int |V_e|(1 - cos(2\omega_{ref}t))dt \\
&\cong -\frac{1}{RC}\frac{|V_{ref}|}{20[V]}\int |V_e|dt
\end{aligned} \qquad (5.15)$$

Like for the previous solution, the low-pass behavior of the integrator minimizes the V_{mul} term at $2\omega_{ref}$ and the demodulation process is achieved without any additional low-pass filter.

Finally, the feedback signal V_{fb} is obtained by modulating V_{ctrl} by means of another AD633:

$$V_{fb} = V_{ref}\left(\frac{1}{10[V]}V_{ctrl} + 1\right) \qquad (5.16)$$

The resulting R_{VCR} value is therefore evaluated as:

$$R_{VCR} = R_b\frac{10[V] + V_{ctrl}}{10[V] - V_{ctrl}} \qquad (5.17)$$

As previously stated, when the system is balanced, V_S equals V_{fb}, and thus, the relationship between V_{ctrl} and x is described by the very simple relation:

$$V_{ctrl} = 10[V]x, \quad \text{i.e., } x = \frac{1}{10[V]}V_{ctrl} \qquad (5.18)$$

Sensitivity $S_{Vctrl,x}$ is constant and equal to 10 V. Its value could be modified acting on the gain of the feedback path.

In terms of readout speed, a possible approach to reduce the settling time of the feedback loop would be to actually decouple the filtering action from the demodulation action [11], as shown in Figure 5.14. By doing this, it is possible to freely tune the loop controller adding also proportional and derivative actions other than the integral one.

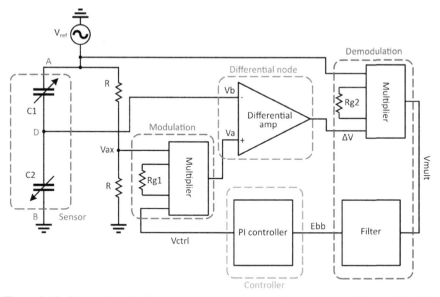

Figure 5.14 Decoupling the filtering action from the demodulator to use a PI controller [11].

5.4 Parasitic Compensation in Autobalanced Bridge-based Differential Capacitive Sensor Interfaces

Figure 5.15(a) recalls the equivalent model of a real-world differential capacitive sensor shown in Chapter 2. There are three main contributions to the overall stray impedances, called C_{pu}, C_p, and C_{pl}.

One of the advantages that inherently comes from an AC-driven setup is the mitigation of the effects of both C_{pu} and C_{pl} [12–16], due to the fact that C_{pl} is fully grounded, while C_{pu} is constantly facing the very low impedance of the signal generator, and hence, its effects are negligible. The compensation method [17] is therefore focused on mitigating only the effects of C_p (see Figure 5.15(b)), which takes a portion of I_1 current (I_p) from I_2 causing a readout error. The basic idea is to have a secondary feedback circuitry (compensation loop, see Figure 5.16) acting in parallel with the main loop analyzed in the previous sections, which, once at the steady state, is capable of compensating the presence of the parasitic capacitance C_p.

The compensation is carried out by suitably driving a voltage-controlled negative impedance converter (VCNIC) which makes sure to deliver to the parasitic capacitance the suitable portion of current ensuring that I_1 is equal

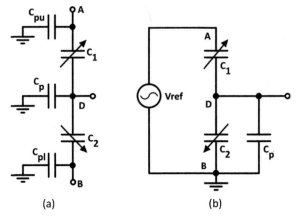

(a) (b)

Figure 5.15 Real-world differential capacitive sensor (a) equivalent model and (b) driven by an AC source.

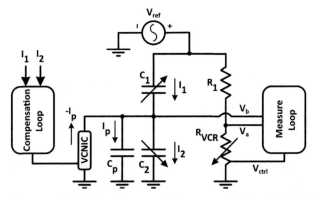

Figure 5.16 Compensation feedback loop applied to an auto-balanced bridge interface [17].

to I_2. The difference between these two currents indeed represents the error signal that the secondary feedback aims to nullify.

The architecture of the negative feedback loop is shown in Figure 5.17. The working principle of the proposed circuit can be summarized as follows. Section A evaluates I_1 and I_2 converting them into proportional voltages and extracting their difference, that is, the high-frequency error signal ($V_{e,hf}$). Supposing to write these two currents as Equation (5.18), the equivalent voltage conversion can be expressed as in Equation (5.20):

$$I_{1,2} = |I_{1,2}| \sin(\omega t) \tag{5.19}$$

$$V_{1,2} = |I_{1,2}| R_{sens} \sin(\omega t) \tag{5.20}$$

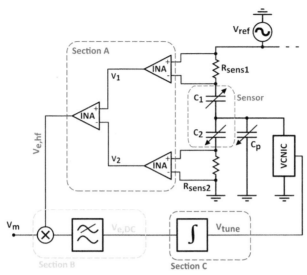

Figure 5.17   The feedback loop schematic diagram.

where $R_{sens} = R_{sens1} = R_{sens2}$. Therefore, the high-frequency error $V_{e,hf}$ can be evaluated as:

$$V_{e,hf} = (|I_1| - |I_2|)R_{sense}\sin(\omega t) \tag{5.21}$$

Section B demodulates the high-frequency error signal extracting the DC component (see Equation (5.22)). This operation is performed by multiplying $V_{e,hf}$ by a known reference voltage, V_m, in phase with I_1 and I_2 and then low-pass filtering the resulting signal.

$$V_{e,DC} = \frac{1}{2}(|I_1| - |I_2|)R_{sense}|V_{ref}| \tag{5.22}$$

Lastly, Section C is composed by the feedback controller. It integrates the DC error producing the *VCNIC* control voltage V_{tune}:

$$V_{tune} = \alpha R_{sense}|V_{ref}|\int_t (|I_1| - |I_2|t)dt \tag{5.23}$$

Once the compensation loop reaches the steady state, V_{tune} can be finally evaluated as:

$$V_{tune} = -10\frac{|C_{par}|}{C_{nic}} + 10 \tag{5.24}$$

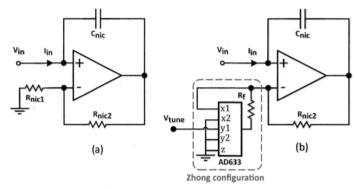

Figure 5.18   (a) The utilized negative impedance converter and (b) the complete VCNIC.

The voltage-controlled negative impedance converter is implemented by means of a modified standard NIC (see Figure 5.18(a)). The input impedance of such a device can be computed as:

$$Z_{in} = \frac{I_{in}}{V_{in}} = -\frac{1}{j\omega \frac{R_{nic2}}{R_{nic1}} C_{nic}} = \frac{1}{j\omega C_{eq}} \tag{5.25}$$

where C_{eq} is:

$$C_{eq} = -\frac{R_{nic2}}{R_{nic1}} C_{nic} \tag{5.26}$$

In order to obtain its voltage-controlled equivalent, it is possible to change the R_{nic2}/R_{nic1} ratio through a suitable VCR obtained by implementing R_{nic1} through a Zhong-connected AD633 as shown in Figure 5.6 (Figure 5.18(b)). Therefore, by substituting Equation (5.3) into (5.26) (substituting V_{ctrl} with the actual driving voltage V_{tune}), it is possible to link the VCNIC equivalent capacitance and the control voltage V_{tune}, as follows:

$$C_{eq} = -R_{nic2} C_{nic} \frac{10[V] - V_{tune}}{10[V] R_f} \tag{5.27}$$

Choosing R_{nic2} equal to R_f, Equation (5.27) becomes:

$$C_{eq} = -C_{nic} \frac{10[V] - V_{tune}}{10[V]} \tag{5.28}$$

As it comes out from this analysis, there is a linear relationship between the negative capacitance and the tuning voltage V_{tune}. The actual span for the VCNIC is dependent both from the feedback capacitance C_{nic} and from the control voltage full-scale.

References

[1] P. Mantenuto, A. De Marcellis and G. Ferri, "Novel modified De-Sauty autobalancing bridge-based analog interfaces for wide-range capacitive sensor applications," *IEEE Sensors Journal*, 14(5), 1664–1672, 2014.

[2] T. Shirakawa, R. Sakai and S. Nakatake, "On-chip impedance evaluation with auto-calibration based on auto-balancing bridge," 2018 IEEE 61st International Midwest Symposium on Circuits and Systems *(MWSCAS)*, Windsor, ON, Canada, 2018, pp. 262–265.

[3] P. Chang and Chih-Ping Liang, "Automatic bridge for inductive voltage dividers," *IEEE Transactions on Instrumentation and Measurement*, 44(2), 418–421, 1995.

[4] A. De Marcellis, G. Ferri and P. Mantenuto, "Automatic analog Wheatstone bridge for wide-range resistive sensor interfacing applications," *Lecture Notes in Electrical Engineering*, 162, 535–539, 2014.

[5] P. Mantenuto, A. De Marcellis and G. Ferri, "Uncalibrated analog bridge-based interface for wide-range resistive sensor estimation," *IEEE Sensors Journal*, 12(5), 1413–1414, 2012.

[6] Ferri G., Parente F.R., Stornelli V., Barile G., Pennazza G. and Santonico M., Voltage-mode analog interfaces for differential capacitance position transducers. In: Andò B., Baldini F., Di Natale C., Marrazza G., Siciliano P. (eds) *Sensors*. CNS 2016. Lecture Notes in Electrical Engineering. Cham: Springer, 2018, Vol. 431.

[7] G. Ferri, V. Stornelli, F. Parente and G. Barile, "Full range analog Wheatstone bridge-based automatic circuit for differential capacitance sensor evaluation," *International Journal of Circuit Theory and Applications*, 45(12), 2149–2156, 2016.

[8] G. Ferri, F. Parente, V. Stornelli, G. Barile and L. Pantoli, "Automatic bridge-based interface for differential capacitive full sensing," *Procedia Engineering*, 168, 1585–1588, 2016.

[9] G. Barile, G. Ferri, F. Parente, V. Stornelli, A. Depari, A. Flammini et al., "A standard CMOS bridge-based analog interface for differential capacitive sensors." In: 2017 13th Conference on Ph.D. Research in Microelectronics and Electronics (PRIME), IEEE, 2017, pp. 281–284.

[10] A. Depari et al., "Autobalancing analog front end for full-range differential capacitive sensing," *IEEE Transactions on Instrumentation and Measurement*, 67(4), 885–893, 2018.

[11] G. Barile et al., "A CMOS full-range linear integrated interface for differential capacitive sensor readout," *Sensors and Actuators A: Physical*, 281, 130–140, 2018.

[12] G. Scotti, S. Pennisi, P. Monsurrò and A. Trifiletti, "88-uA 1-MHz stray-insensitive CMOS current-mode interface IC for differential capacitive sensors," *IEEE Transactions on Circuits and Systems I: Regular Papers*, 61(7), 1905–1916, 2014.

[13] S. Ogawa, "A low-power, high-accuracy capacitance-to-time converter for differential capacitive sensors," 2017 IEEE 8th Latin American Symposium on Circuits & Systems (LASCAS), Bariloche, 2017, pp. 1–4.

[14] F. Reverter and Ò. Casas, "Interfacing differential capacitive sensors to microcontrollers: A direct approach," *IEEE Transactions on Instrumentation and Measurement*, 59(10), 2763–2769, 2010.

[15] S. Malik, K. Kishore, T. Islam, Z. Zargar and S. A. Akbar, "A time domain bridge-based impedance measurement technique for wide range lossy capacitive sensors," *Sensors and Actuators A: Physical*, 234, 248–262, 2015.

[16] S. Baglio, S. Castorina, G. Ganci and N. Savalli, "A high sensitivity conditioning circuit for capacitive sensors including stray effects compensation and dummy sensors approach," Proceedings of the 21st IEEE Instrumentation and Measurement Technology Conference (IEEE Cat. No. 04CH37510), Como, 2004, pp. 1542–1545, Vol. 2.

[17] G. Barile et al., "Fully analog automatic stray compensation for bridge-based differential capacitive sensor interfaces," 2018 International Conference on IC Design & Technology (ICICDT), Otranto, Italy, 2018.

Appendix A: Second-generation Current and Voltage Conveyors: A Short Review

This appendix aims to give to the reader a deeper overview of the second-generation current conveyor (CCII) and of its dual version, the second-generation voltage conveyor (VCII). Both of them have been used as active device in the interfaces introduced along the text.

The analysis will be carried out both from a high-level point of view, evidencing the nonidealities of these devices with respect to their ideal use cases, and from a transistor-level point of view, showing both basic and more advanced architectures.

A.1 Second-generation Current Conveyor

As the name suggests, the CCII was conceived in 1970 as the successor of the previously proposed first-generation current conveyor (CCI) with the aim of improving its flexibility by adding a voltage input terminal [1–3]. Indeed, a CCII (Figure A.1) is a three-terminal active device having a high-impedance voltage input (Y), a low-impedance current input (X), and a high impedance current output (Z). Ideally, the relationships that link each terminal can be expressed by the following matrix:

$$\begin{bmatrix} I_Y \\ V_X \\ I_Z \end{bmatrix} = \begin{bmatrix} 0 & 0 & 0 \\ 1 & 0 & 0 \\ 0 & \pm 1 & 0 \end{bmatrix} \begin{bmatrix} V_Y \\ I_X \\ V_Z \end{bmatrix} \tag{A.1}$$

From Equation (A.1), it comes out that a voltage that is applied to the Y terminal is ideally buffered to the X node, whereas a current that is generated at the X terminal is "conveyed" to the output Z port. The direction of the current vector is determined by the CCII itself. As visible from Figure A.1(a), in a positive current conveyor (CCII+), currents at X and Z share the same direction with respect to the device terminals. On the other hand, as shown in from Figure A.1(b), in a negative current conveyor (CCII−), the same

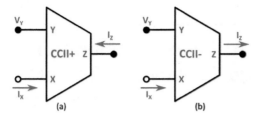

Figure A.1   High level representation of a (a) CCII+ and (b) CCII−.

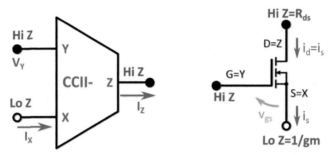

Figure A.2   Analogy between a CCII and a MOSFET.

currents have opposite directions. Moreover, being Y a high (ideally infinite) impedance node, no current flows in it whenever a voltage is applied.

From the matrix reported in Equation (A.1), it is remarkable that the analogy between a CCII and a MOSFET. Indeed, as shown in Figure A.2, the buffering action between the Y and X terminal resembles the behavior of the gate-source voltage, being the gate a high-impedance input and the source a low-impedance voltage output (given by the inverse of the transistor transconductance *1/gm*). Similarly, the drain current matches the source current, mimicking the conveying action between the low-impedance current input X (the source of the MOSFET) and the high-impedance current output Z (the drain of the MOSFET, being equal to the drain-source resistance R_{ds}).

Compared to the widespread Op-Amp, a CCII, thanks to its native current and voltage processing capability, allows to achieve a wider bandwidth given a fixed gain level and allows to reach better common mode rejection ratios in many applications like instrumentation amplifiers [4–7].

A.1.1 CCII Nonidealities

The nonidealities of a CCII reside in three main aspects: the non-unitary voltage and current buffering between the terminals, their constant voltage

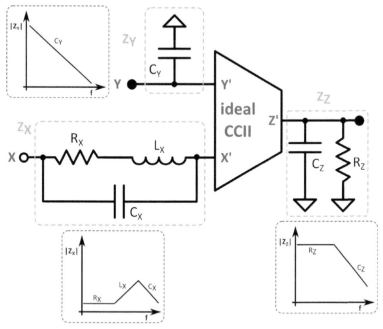

Figure A.3   CCII impedances at each terminal.

or current offsets, and, most importantly, the impedance levels at the same X, Y and Z terminals. Indeed, a more realistic version of Equation (A.1) is given in the following Equation (A.2):

$$
\begin{bmatrix} I_Y \\ V_X \\ I_Z \end{bmatrix} = \begin{bmatrix} sC_Y & 0 & 0 \\ \alpha & (R_X + sL_X)//\dfrac{1}{sC_X} & 0 \\ 0 & \pm\beta & \dfrac{1}{R_Z//\dfrac{1}{sC_Z}} \end{bmatrix} \begin{bmatrix} V_Y \\ I_X \\ V_Z \end{bmatrix} \quad \text{(A.2)}
$$

The elements on the main diagonal represent the actual impedance (or admittance) at each port (see Figure A.3). As visible, at Y terminal the impedance is purely capacitive: this is a typical feature of CCII designed using CMOS differential pair inputs. A resistive component may be present in the case of BJT input stage implementations or even in the case of CMOS stages where the input is different from a gate terminal. Since ideally the impedance level at Y is infinite, as the operating frequency increases, the CCII

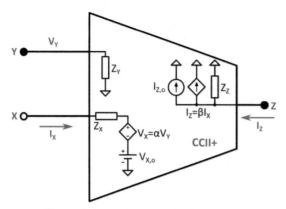

Figure A.4   Real CCII equivalent circuit.

loses ideality. At X terminal, the main low-frequency contribution is typically resistive and very low in magnitude. As the frequency increases, the inductive contribution becomes predominant increasing the overall impedance at this terminal. Similarly, at Z output terminal, the low-frequency impedance is purely resistive with a large magnitude. At high frequency, however, the capacitive contribution lowers it. The bandwidth of the impedance (identified as the frequency interval where the value of the impedance remains acceptable for a specific application) is an important metric to evaluate the ideality of a CCII.

The α and β parameters represent the voltage and current buffering coefficients, respectively. Their magnitude is ideally equal to unity, although in some modified versions of the CCII [8, 9] they are made greater than one for a specific purpose.

A real equivalent circuit of the CCII is represented as in Figure A.4. As visible, this model takes into account the impedances at each port, α and β parameters, and also voltage and current offsets at X ($V_{X,o}$) and Z ($I_{Z,o}$) terminals.

A.1.2 CCII Topologies

There are many possible topologies that implement, at transistor level, a CCII [10–14]. We will analyze solutions based on a classical differential pair because we want to show how, starting from a very simple structure, it is possible to improve it maintaining the same approach thus obtaining more performing and advanced solutions.

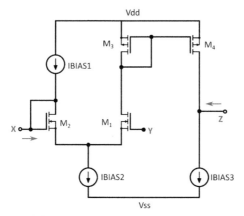

Figure A.5 Class A CCII based on a differential pair.

A basic topology that implements a current conveyor is shown in Figure A.5 (please consider the scheme as purely ideal, in particular concerning the current sources). Transistors M_1 and M_2 implement the differential pair. A simple unitary negative feedback at M_2 creates the low-impedance current input X, whereas M_3 and M_4 make sure that the current that is forced into X is mirrored at the high-impedance Z terminal.

A simple small signal analysis allows to evaluate both α and β parameters, and CCII impedances. The former parameters can be evaluated as:

$$\alpha = \frac{R_{ds1}gm_1}{1 + R_{ds2}gm_2} \cong \frac{gm_1}{gm_2} \tag{A.3}$$

$$\beta = \frac{R_{ds3}gm_4}{1 + R_{ds3}gm_3} \cong \frac{gm_4}{gm_3} \tag{A.4}$$

If M_1 and M_2 have the same dimensions, α results to be close to unity. Unlike this, β depends on the load impedance, and therefore, the approximation of Equation (A.4) stands only if it is negligible compared to R_{ds4}.

Terminal impedances can be evaluated, respectively, as:

$$Z_X \cong \frac{1}{gm_2} \tag{A.5}$$

$$Z_Y = \frac{1}{(WL)_{M1}C_{ox}} \tag{A.6}$$

$$Z_Z = \frac{R_{ds4}R_{bias3}}{R_{ds4} + R_{bias3}} \tag{A.7}$$

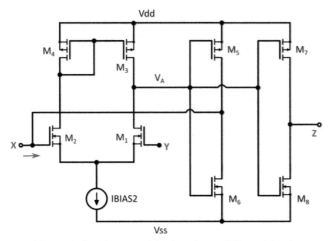

Figure A.6   Class AB CCII based on a differential pair.

Concerning low and medium frequencies, the high-input impedance at the Y terminal depends only on the geometry of M_1 (and on the technology used), whereas the low impedance at the X port is given by the inverse of the transconductance of M_2 transistor, being it diode-connected. The high impedance at the output Z port can be evaluated as the parallel between M_4 and *IBIAS3* output resistances.

An upgraded class AB CCII, based on a differential input pair, is shown in Figure A.6. As visible, the CCII is composed by three stages, a simple operational transconductance amplifier (OTA) as input stage $(M_1 - M_4)$, and two CMOS inverters $(M_5, M_6; M_7, M_8)$ working in parallel.

Transistors M_5 and M_6 provide the necessary signal inversion that performs the negative feedback generating the low impedance X terminal; M_7 and M_8, if made equal to M_5 and M_6, respectively, make sure that the current flowing through the X terminal is suitably conveyed to Z terminal.

Since transistors $M_1 - M_6$ form a dual stage Miller OTA, stability has to be studied in order to safely close the loop. The open loop gain of the Miller OTA can be calculated as:

$$A_V = gm_1(R_{ds1}//R_{ds3})(gm_6 + gm_5)(R_{ds6}//R_{ds5}) \qquad (A.8)$$

The higher the A_v is, the lower the low-frequency input impedance at the X terminal:

$$R_X = \frac{R_{ds6}//R_{ds5}}{1 + A_V} = \frac{R_{ds7}//R_{ds8}}{1 + A_V} = \frac{R_Z}{1 + A_V} \qquad (A.9)$$

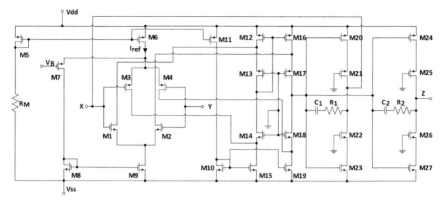

Figure A.7 RtR constant gm CCII based on a differential pair.

having designed $M_5 = M_7$ and $M_6 = M_8$. The Y terminal impedance is the same as Equation (A.6). Similarly, the parallel between the drain-source resistances of M_7 and M_8 is the equivalent low-frequency impedance at Z node.

Based on the same differential pair concept, Figure A.7 shows a rail-to-rail CCII topology [15].

In order to achieve the full swing at the input, n-MOS (M_1, M_2) and p-MOS (M_3, M_4) differential pairs are placed in parallel, so as to proportionally activate one or the other, according to the input amplitude and common mode.

To make sure that the CCII performs equally regardless of the previously mentioned parameters, the circuit is also capable of maintaining a constant total gain gm_T. To do that, input transistors can be biased in weak inversion region, so that the total transconductance results proportional to the current:

$$gm_T = \frac{I_P}{2n_p\,V_T} + \frac{I_N}{2n_n\,V_T} \qquad (A.10)$$

where I_P is the biasing current flowing in the input p-channel pair and I_N is the same for the input n-channel pair and n is an index ranging from 1 to 2.

The constant-g_m condition is established through the switch M_7 and the current mirror M_8, M_9. When a low common mode input is applied, the reference current (I_{ref}), set through R_M, M_5, M_6, flows only into the input p-type differential pair, being M_7 off. If, on the other hand, the common mode voltage increases (see Figure A.7), the current switch M_7 allows I_{ref} to flow through M_8 and M_9 transistors and into the n-type input differential pair. The sum of currents flowing in both complementary pairs is always kept equal to I_{ref}, and so it remains the same at any common mode level. The value of V_B determines the range when one differential pair is on and the other is off.

In this topology, it is also shown how it is possible to use a different OTA topology (folded cascode) than a Miller one. The second stage ($M_{20} - M_{23}$) and the matched third stage ($M_{24} - M_{27}$) are developed as cascoded class-AB inverters to increase Z impedance. The low-frequency impedance at Y port results to be:

$$Z_Y = Z_{M2}//Z_{M4} \tag{A.11}$$

where Z_{M2} and Z_{M4} can be calculated as in Equation (A.6). By choosing a cascoded inverter as second stage, the impedance at X node is reduced, since the gain of the open-loop equivalent OTA composed by the folded cascode $M_1 - M_{19}$ and the cascoded inverter $M_{20} - M_{23}$ is increased and equal to:

$$A_V = A_I A_{II} \tag{A.12}$$

where A_I corresponds to the folded cascode input stage gain and can be evaluated as:

$$A_I = G_{MI}\{[g_{m18}r_{ds18}(r_{ds19}//r_{ds4})]//[g_{m17}r_{ds17}(r_{ds16}//r_{ds2})]\} \tag{A.13}$$

and A_{II} is the gain of the cascode inverter second stage and can be calculated as:

$$A_{II} = G_{MII}[(g_{m21}r_{ds20}r_{ds21})//(g_{m22}r_{ds23}r_{ds22})] \tag{A.14}$$

In Equations (A.13) and (A.14), the terms G_{MI} and G_{MII} can be evaluated as:

$$G_{MI} = g_{mT} = g_{mP} + g_{mN}; \quad G_{MII} = g_{m20} + g_{m23} \tag{A.15}$$

The output impedance at the Z node, for low frequencies, corresponds to the impedance of a folded cascode and can be evaluated as:

$$Z_Z = (g_{m25}r_{ds24}r_{ds25})//(g_{m26}r_{ds27}r_{ds26}) \tag{A.16}$$

A.2 Second-generation Voltage Conveyor

The second-generation voltage conveyor (VCII) was born as the dual block of the CCII, which inherits the "second-generation" appellation, even though a

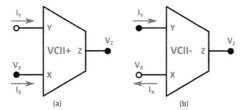

Figure A.8 High level representation of a (a) VCII+ and (b) VCII−.

first-generation voltage conveyor does not really exist. The idea of theorizing such an active device dates back to two decades ago [16, 17]; however, only recently, its capabilities are gaining scientific attention and transistor level synthesis are proposed [18, 20, 24–27].

As shown in Figure A.8, where a generic voltage conveyor symbol is reported, it is still a three-port block, where terminals are denoted, again, by X, Y, and Z. Unlike CCII, however, Y and Z terminals show a low impedance (ideally zero), being a current input and a voltage output terminal, respectively, while X presents a high impedance (ideally infinite) being a voltage input terminal. The ideal outputs of its three ports in terms of their corresponding inputs are shown in the matrix of Equation (A.17):

$$\begin{bmatrix} i_x \\ v_y \\ v_z \end{bmatrix} = \begin{bmatrix} 0 & \pm1 & 0 \\ 0 & 0 & 0 \\ 1 & 0 & 0 \end{bmatrix} \begin{bmatrix} v_x \\ i_y \\ i_z \end{bmatrix} \tag{A.17}$$

From the last equation, it comes that a VCII consists of a current buffer between Y and X terminals and a voltage buffer between X and Z ones.

By therefore forcing a current to flow into the Y node, it is mirrored to the X node, whereas, by setting a voltage at the X port, it is buffered to the Z port. Ideally, Y terminal can be considered at virtual ground.

Although very simple, VCII has a great potential, since it joins both the benefits of Op-Amps and CCIIs. In fact, it is able to work both in a voltage mode and in a current mode environment, as well as to create a linking point between them.

Op-Amp-based circuits are typically limited to produce inverting type outputs. On the contrary, by replacing Op-Amps with VCIIs in these applications, both inverting and non-inverting outputs can be easily produced without modifying the circuital configuration, but only the internal topology. Also, VCIIs are able to overcome problems such as current crosstalk in non-inverting voltage summing amplifiers which, for an Op-Amp, cannot be avoided given that only high-impedance inputs are available. Another crucial

feature of the VCII is the wide and gain-independent bandwidth, whereas, for
Op-Amp-based counterparts, bandwidth reduces proportionally by increasing
gain [19].

Since the emergence of current-mode signal processing, where it is possible to achieve fast and reliable signal processing with a very low voltage
level (condition that perfectly suits for the new scaled technologies), CCIIs
can replace Op-Amp in many applications. Indeed, as shown in the previous
paragraph, they have a very simple internal topology and work in open-
loop conditions. These things make the CCII easier to utilize and able to
reach wider bandwidths. In spite of these advantages, the lack of a low
impedance voltage output port forces designers to add extra voltage buffers
in applications requiring a voltage output. VCIIs deal with this limitation
offering a low-impedance voltage output.

A.2.1 VCII Nonidealities

Equation (A.18) shows a real-world input–output relationship matrix for
a VCII:

$$
\begin{bmatrix} I_X \\ V_Y \\ V_Z \end{bmatrix} =
\begin{bmatrix} \dfrac{1}{r_x} + sC_x & \pm\beta & 0 \\ 0 & r_y + sL_y & 0 \\ \alpha & 0 & r_z + sL_z \end{bmatrix}
\begin{bmatrix} V_X \\ I_Y \\ I_Z \end{bmatrix} \tag{A.18}
$$

The elements on the main diagonal represent the equivalent impedance
or admittance at each terminal (see Figure A.9). Indeed, r_y, L_y, r_x, C_x, r_z
and L_z are the parasitic resistances associated with Y terminal (ideally $= 0$),
the parasitic inductance at Y terminal (ideally $= 0$), the parasitic resistance at
X terminal (ideally $= \alpha$), the parasitic capacitance at X terminal (ideally $= 0$),

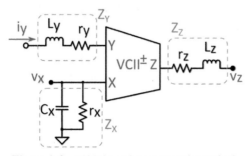

Figure A.9   VCII impedances at each terminal.

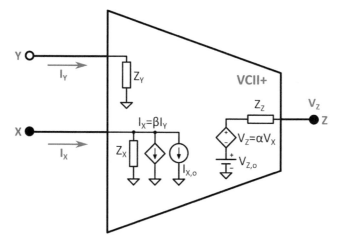

Figure A.10 Real VCII equivalent circuit.

the parasitic resistance at Z terminal (ideally $= 0$), and the parasitic inductance at Z terminal (ideally $= 0$), respectively. Coefficients α and β have the same meaning as for a CCII, with the former the ratio between voltages at Z and X, whereas the latter referencing to the ratio between currents at X and Y (sign denotes whether currents are share the same or opposite direction).

Figure A.10 depicts a real-world equivalent model for a VCII. Similar to the CCII, other than the effects of impedances and mirroring coefficients, offset voltages and currents have been represented as well.

A.2.2 VCII Topologies

In this section, a couple of CMOS-based VCII topologies are introduced. They have particular features to be used in different application fields.

The first architecture is shown in Figure A.11 [20]. Like previously stated, it is made of a current buffer between Y and X terminals and a voltage buffer between X and Z terminals.

The former is made of transistors $M_1 - M_7$ and the current sources $I_{B1} - I_{B4}$, whereas the latter is implemented by M_8, $M_{A1} - M_{A3}$ and $I_{B5} - I_{B7}$ current sources. A first negative feedback loop, established by $M_1 - M_3$, forces the offset voltage at Y terminal to be equal to ground, lowering its impedance. The second negative feedback loop, formed by $M_4 - M_7$ transistors, has the dual task to further reduce the impedance at Y terminal while mirroring the input current to the X terminal. Based on the small signal

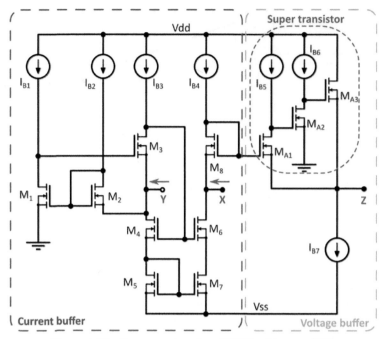

Figure A.11   Super transistor-based VCII architecture [20].

equivalent circuit (see Figure A.12), the low-frequency impedance at Y and X terminals can be calculated as:

$$r_Y = \frac{1}{gm_3 gm_1 (ro_1 \| ro_{IB_1}) gm'_4 (ro_{IB_3} \| ro_3)} \tag{A.19}$$

$$r_X = \frac{(ro_7 + ro_6 + gm_6 ro_6 ro_7)(ro_{IB_4} + 1/gm_8)}{(ro_7 + ro_6 + gm_6 ro_6 ro_7 + ro_{IB_4} + 1/gm_8)} \tag{A.20}$$

Based on the same model, it is possible to extract the β equation:

$$\beta = \frac{i_x}{i_y} = \frac{gm'_6 ro_{IB_3}(R_{eq} \| ro_{IB_2})}{1 + gm'_4 ro_{IB_3}(R_{eq} \| ro_{IB_2})} \approx \frac{gm'_6}{gm'_4} = \frac{gm_7}{gm_5} \tag{A.21}$$

where:

$$gm'_4 = \frac{gm_4 gm_5}{gm_5 + gm_4}, \quad gm'_6 = \frac{gm_4 gm_7}{gm_5 + gm_4} \tag{A.22}$$

Parameters gm_i and ro_i represent the transconductance and the output impedance of the related transistor, respectively, and ro_{IBi} is the output impedance of related current source, while R_{eq} is the equivalent resistance seen from the drain of M_4.

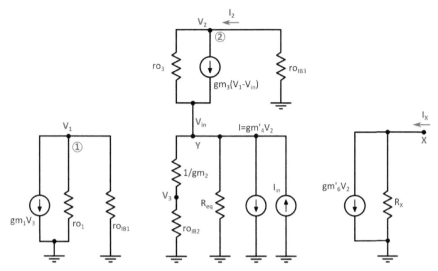

Figure A.12 Small signal equivalent circuit for the X and Y terminals of the VCII under analysis.

To implement the Z stage of the VCII, the presented topology uses a super transistor $(M_{A1} - M_{A3})$ [21] which, thanks to its negative feedback, provides a very low impedance at Z terminal also improving the buffering between X and Z terminals. A small signal analysis of this structure [21] allows to derive the voltage transfer between X and Z terminals and the Z terminal impedance, respectively:

$$\alpha = \frac{v_z}{v_x}$$

$$= \frac{ro_{IB4}}{ro_{IB4} + gm_8{}^{-1}} \frac{1}{1 + [gm_{A1}gm_{A2}(ro_{A1}\|ro_{IB5})(ro_{A2}\|ro_{IB6})]^{-1}}$$

(A.23)

$$r_Z = \frac{1}{gm_{A1}gm_{A2}gm_{A3}(ro_{A1}\|ro_{IB5})(ro_{A2}\|ro_{IB6})}$$

(A.24)

Figure A.13 shows a flipped voltage follower [22, 23] based VCII [24–27]. The Y branch of the VCII is implemented by means of an AB-class super common-gate cell formed by M_1 – M_6 together with current sources I_{B1} and I_{B2}. The local negative feedback performed by the differential pair M_1, M_2 and M_5 fixes to ground the voltage at Y helping to lower the impedance at the same node. Transistors M_6 and M_7 implement a simple

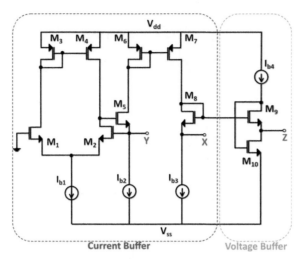

Figure A.13   Flipped voltage follower-based VCII architecture.

current mirror to make sure that the current forced into the Y terminal is suitably mirrored to the X terminal.

Let us consider now the expressions of low-frequency impedances at VCII terminals. The low-input impedance at the Y port can therefore be approximatively evaluated as:

$$r_Y \cong \frac{1}{gm_5 gm_2 (R_{ds2} // R_{ds4})} \qquad (A.25)$$

The high impedance at the X node is ensured by the presence of the drain of M_7 together with the current source I_{B3} (which can be implemented with a simple current mirror). Indeed, it can be calculated as:

$$r_X = \frac{R_{ds7} + R_{ds8} + gm_8 R_{ds8} R_{ds7}}{1 + gm_8 R_{ds8}} // R_{OIb3} \cong R_{ds7} // R_{OIb3} \quad (A.26)$$

M_8 acts as a level shifter to properly adjust the DC value at the input of the FVF voltage buffer. Since it constitutes the Z output node, the impedance at this terminal can be easily evaluated as:

$$r_Z \cong \frac{1}{gm_9 gm_{10} R_{ds9}} \qquad (A.27)$$

References

[1] C. K. Smith and A. Sedra, "The current conveyor – new building block," *IEEE Proceedings*, 56, 1368–1369, 1968.

[2] A. Sedra and C. K. Smith, "A second-generation current conveyor and its applications," *IEEE Transactions on Circuit Theory*, CT-17, pp.132–134, 1970.

[3] A. S. Sedra, G. W. Roberts, I. Gohh, The current conveyor: History, progress and new results. *IEEE Proceedings*, 137, Pt. G, No. 2, 78–87, 1990.

[4] S. Maheshwari, "High CMRR wide bandwidth instrumentation amplifier using current controlled conveyors," *International Journal of Electronics*, 91(2), 137–137, 2004.

[5] S. Mago, H. Tamura and K. Tanno, "High CMRR and wideband current feedback instrumentation amplifier using current conveyors," Proceedings of International Conference on Artificial Life and Robotics, 22, 532–535, 2017.

[6] A. Hou, "A wide bandwidth isolation amplifier design using current conveyors," *Analog Integrated Circuits and Signal Processing*, 40(1), 31–38, 2004.

[7] H. Ercan, S. A. Tekin and M. Alçı, Voltage- and current-controlled high CMRR instrumentation amplifier using CMOS current conveyors. *Turkish Journal of Electrical Engineering and Computer Sciences*, 2012.

[8] W. Surakampontorn and K. Kumwachara, "CMOS-based electronically tunable current conveyor," *Electronics Letters*, 28(14), 1316–1317, 1992.

[9] A. Fabre and N. Mimeche, "Class A/AB second-generation current conveyor with controlled current gain," *Electronics Letters*, 30(16), 1267–1269, 1994.

[10] G. Ferri, N. C. Guerrini, "Low-voltage low-power CMOS current conveyors." Springer, 2003. ISBN 9780306487200.

[11] G. Kapur, S. Mittal, C. M. Markan and V. P. Pyara, "Design of analog field programmable CMOS current conveyor," *Science Journal of Circuits, Systems and Signal Processing*, 1(1), 9–21, 2012.

[12] F. Khateb, S. Bay Abo Dabbous and S. Vlassis, "A survey of non-conventional techniques for low-voltage low-power analog circuit design," *Radioengineering*, 22, 415–427, 2013.

[13] Y. Hwang, Y. T. Ku, J. Chen and C. Yu, "Inverter-based low-voltage CCII design and its filter application," *Radioengineering*, 22, 1026–1033.

[14] T. Ettaghzouti, N. Hassen and K. Besbes, "A novel low-voltage low-power CCII based on super class AB CMOS OTA cells and filter application," 12th International Multi-Conference on Systems, Signals & Devices, Sakiet Ezzit Sfax, March 2015.

[15] V. Stornelli, G. Ferri, L. Pantoli, G. Barile and S. Pennisi, "A rail-to-rail constant-g_m CCII for instrumentation amplifier applications," *AEU – International Journal of Electronics and Communications*, 91, 103–109, 2018.

[16] I. M. Filanovsky, "Current conveyor, voltage conveyor, gyrator." Proceedings of the 44th IEEE 2001 Midwest Symposium on Circuits and Systems, 1, 314–317, 2001.

[17] J. Cajka and K. Vrba, "The Voltage Conveyor May Have in Fact Found its Way into Circuit Theory," *International Journal of Electronics and Communications (AEUE)*, 58, 244–248, 2004.

[18] J. Koton, N. Herencsar and K. Vrba, "KHN-equivalent voltage-mode filters using universal voltage conveyors," *International Journal of Electronics and Communications (AEUE)*, 65, 154–160, 2011.

[19] G. Barile, L. Safari, G. Ferri and V. Stornelli, "Traditional Op-Amp and new VCII: A comparison on analog circuits applications," *AEU – International Journal of Electronics and Communications*, 152845, 2019.

[20] L. Safari, G. Barile, V. Stornelli and G. Ferri, "An overview on the second generation voltage conveyor: Features, design and applications," *IEEE Transactions on Circuits and Systems II: Express Briefs*, 66(4), 547–551, 2019.

[21] L. Safari and S. Azhari, "A novel wide band super transistor based voltage feedback current amplifier," *AEU – International Journal of Electronics and Communications*, 67(7), 624–631, 2013.

[22] R. Carvajal, J. Ramirez-Angulo, A. Lopez-Martin, A. Torralba, J. Galan, A. Carlosena and F. Chavero, "The flipped voltage follower: A useful cell for low-voltage low-power circuit design," *IEEE Transactions on Circuits and Systems I: Regular Papers*, 52(7), 1276–1291, 2005.

[23] F. Centurelli, P. Monsurrò, D. Ruscio and A. Trifiletti, "A new class-AB flipped voltage follower using a common-gate auxiliary amplifier," 2016 MIXDES – 23rd International Conference Mixed Design of Integrated Circuits and Systems, Lodz, Poland, 2018, pp. 143–146.

[24] L. Safari, G. Barile, V. Stornelli, G. Ferri and A. Leoni, "New current mode Wheatstone bridge topologies with intrinsic linearity," 2018 14th Conference on Ph.D. Research in Microelectronics and Electronics (PRIME), Prague, Czech Republic, 2018, pp. 9–12.

[25] G. Barile, A. Leoni, L. Pantoli, L. Safari and V. Stornelli, "A new VCII based low-power low-voltage front-end for silicon photomultipliers," 2018 3rd International Conference on Smart and Sustainable Technologies (SpliTech), Split, Croatia, 2018.

[26] L. Pantoli, G. Barile, A. Leoni, M. Muttillo and V. Stornelli, "A novel electronic interface for micromachined Si-based photomultipliers," *Micromachines*, 9(10), 507, 2018.

[27] L. Safari, G. Barile, G. Ferri and V. Stornelli, "High performance voltage output filter realizations using second generation voltage conveyor," *International Journal of RF and Microwave Computer-Aided Engineering*, 28(9), e21534, 2018.

Index

About the Authors

Gianluca Barile was born in Avezzano, Italy. He received the master's degree (cum laude) in Electronics Engineering from the University of L'Aquila (Italy) in 2016 and the PhD degree from the same University in 2020. His research activity includes voltage and current mode sensor interfaces, integrated circuits design and system for industrial electronics.

Giuseppe Ferri was born in L'Aquila, Italy, in 1965. He received the "Laurea" degree (cum laude) in electronic engineering in 1988. In 1991, he joined the Department of Electronic Engineering, University of L'Aquila, L'Aquila, Italy, where he is actually a full professor of Electronics and Microelectronics at the University of L'Aquila, Italy. His research activity mainly concerns the design of analog electronic circuits for integrated sensor applications both in voltage and in current-mode. In this field of research, he is author or coauthor of 7 patents, 3 international books, 1 book chapter and more than 400 publications in international journals and conference proceedings. He is PhD coordinator for his department and responsible of the sensors, microsystems and instrumentation area in Italian Electronic Group. He is also an IEEE senior member and Editor of some international Journals and Special Issues.

Vincenzo Stornelli was born in Avezzano, Italy. He received the "Laurea" degree (cum laude) in electronic engineering in 2004. In October 2004, he joined the Department of Electronic Engineering, University of L'Aquila, L'Aquila, Italy, where he is actually involved as Associate Professor with problems concerning current mode applications; physics-based simulation; computer-aided design modeling characterization and design analysis of active microwave components, circuits and subsystems; design of integrated circuits for Sensor and RF applications. He is an IEEE senior member and Editor of several international Journals and Special Issues. He also serves as a reviewer for several international journals.